FOCUS 18 • INTRODUCTION • JULY 2010

Most rural households lack access to reliable and affordable finance for agriculture and other livelihood activities. Many small farmers live in remote areas where retail banking is limited and production risks are high. The recent financial crisis has made the provision of credit even tighter and the need to explore innovative approaches to rural and agricultural finance even more urgent.

Rural and agricultural finance innovations have significant potential to improve the livelihoods and food security of the poor. Although microfinance has been widely studied, a large knowledge gap still exists on the nuts and bolts of expanding access to rural and agricultural finance. IFPRI's 2020 Vision Initiative approached the rural finance team of the Agriculture and Rural Development Department of the World Bank to conceptualize and assemble this collection of briefs to narrow the knowledge gap by examining innovations in providing financial services to rural households. They, in turn, asked leading experts around the world to share their perspectives and experiences, focusing on issues related to implementation and operations. Together with a companion set of briefs—*Innovations in Insuring the Poor* (2020 Focus 17) edited by Ruth Vargas Hill and Maximo Torero—this series contributes to the knowledge pool on innovative tools for effectively managing the risks faced by the rural poor.

This set of briefs clearly points out the importance of business realities faced by small farmers, including low education levels, the dominance of subsistence farming, and the lack of access to modern financial instruments. These conditions mean that new and innovative institutions are required to reach small farmers. Emerging communication technologies provide new opportunities for rural banking by reducing business costs and alleviating information asymmetries. New financing instruments, such as weather index-based insurance and microinsurance, also have great potential for managing the risks faced by small farmers. In addition, bundling financial services with nonfinancial services like marketing and extension services offers new opportunities for small farmers to increase their productivity and incomes. Finally, an enabling policy environment and legal framework, enforcement of rules and regulations, and a supportive rural infrastructure all contribute immensely to making sustainable access to finance a reality.

We are grateful to Renate Kloeppinger-Todd and Manohar Sharma for their work in bringing together these important briefs, to the brief authors for their analyses and insights, to the reviewers for their constructive comments, to Heidi Fritschel and Ashley St. Thomas for editorial assistance, to Shirong Gao for design, and to Djhoanna Cruz for coordination assistance. We hope that the findings and recommendations presented here will contribute to policy changes that enhance poor people's access to financial services in ways that increase their livelihoods and improve their lives.

Shenggen Fan
Director General
IFPRI

Juergen Voegele
Director, Agriculture and Rural Development
World Bank

Rajul Pandya-Lorch
Head, 2020 Vision Initiative
IFPRI

DOI: 10.2499/0896296687

ISBN 10-digit: 0-89629-668-7
ISBN 13-digit: 978-0-89629-668-8

Overview

RENATE KLOEPPINGER-TODD AND MANOHAR SHARMA

FOCUS 18 • BRIEF 1 • JULY 2010

Everywhere in the world, small agricultural producers are entrepreneurs, traders, investors, and consumers, all rolled into one. In all these roles, small agricultural households constantly seek to use available financial instruments to improve their productivity and secure the best possible consumption and investment choices for their families. But the package of financial services available to small farmers in developing countries is severely limited, especially for those living in remote areas with no access to basic market infrastructure.

When poor people have limited saving or borrowing options, their investment plans are stifled and it becomes harder for them to break out of poverty. If households have no access to insurance and are unable to accumulate small savings that enable them to pay for household and business expenses, especially during lean seasons, they are forced to limit their exposure to risk, even if high returns are expected, once again making the pathway out of poverty more arduous than necessary. Inadequate access to financial services is thus part of what is often called the "poverty trap."

Microfinance and agriculture finance

In the 1980s and 1990s the deleterious impact of limited financial access caught the attention of many academics, policymakers, donor agencies, and development practitioners, who generated an outpouring of new thinking and new ideas. Innovative concepts such as group liability, village banking, microinsurance, and index-based insurance were tested in new and emerging microfinance institutions. But progress on expanding agricultural finance—as opposed to nonagricultural microenterprise finance—lagged. Donors and governments that had invested heavily in agricultural development banks and agricultural credit in the 1980s and early 1990s found that these efforts did not produce the expected results and withdrew their support. It was hoped that private commercial banks would step in, but for the most part they did not.

Financial institutions have demonstrated a lack of interest in agriculture finance for four reasons. First, many agricultural households were located in remote parts of the country and were often so widely dispersed that financial institutions found it challenging to provide cost-effective and affordable services. Second, big swaths of the agricultural population were subject to the same weather and climate risks, making it hard for providers of financial services to hedge risks or operate profitable insurance pools. Third, service providers, mainly urban-based, simply did not know enough about the business of agriculture to devise profitable financial products. Fourth, most small agricultural producers in developing countries had little education and little knowledge of how modern banking institutions work.

Recent progress in rural finance

Since the early 2000s a number of organizations have developed innovative approaches to financing agriculture. They have sometimes adapted microfinance concepts to the provision of agricultural finance, used good banking practices, and above all drawn on knowledge of agriculture to enter and succeed in this market. Many of these new approaches show great promise, but no single approach works for all situations. Rather, organizations have the most success when they are nondogmatic, apply comprehensive risk-management strategies and tools, retain the ability to pick and choose their clients rather than having the government do so, and are innovative and pragmatic.

This set of briefs explores how rural and agricultural finance can be profitable, without high levels of government subsidies, by examining a selection of successful interventions—out of the many being implemented in the developing world—and highlighting the lessons learned.

The briefs fall into four thematic areas: addressing the business reality of small farmers in developing countries, using modern communication technology to overcome the tyranny of distance and information bottlenecks, managing risks at the farm and household level, and bundling financial services with nonfinancial services to address the multiple constraints faced by most small farmers.

Addressing the business reality of small farmers

Most small farmers in developing countries have little education and limited exposure to modern financial instruments. Further, many of these small farmers have only recently transitioned from subsistence to commercial farming, and their contact with the cash economy and experience in cash management is limited. Hence, in Brief 2 Monique Cohen addresses the issue of financial literacy and explains why the poor may need some coaching on how modern financial instruments can better their lives.

Additionally, many small farmers in developing countries live in remote rural settings, where urban-based retail banking is unavailable. In Brief 3 Anne Ritchie describes two operational models used by community-based financial organizations and explains how community banking enables the unbanked rural poor to serve themselves, with or without links to the formal financial sector.

As rural banking takes hold in developing countries, it has also attracted the attention of institutions in developed countries that have traditionally served farmers. The Netherlands-based Rabobank, for example, has made investments in countries as varied as China, Paraguay, and Zambia. In Brief 4 Gerard van Empel describes Rabobank's use of a supply-chain approach to address key gaps in rural banking in many developing-country contexts.

Ghana's network of rural and community banks represents a unique approach to generating access to financial services across the rural areas of a whole country. In Brief 5 Ajai Nair and Azeb Fissha describe their business model, their services, and their financial performance. The brief discusses the challenges facing the network and its apex institution in becoming financially sustainable and competitive and draws lessons that are applicable elsewhere.

The financing of productive assets requires access to medium-term loans and usually significant collateral, neither of which are available in most rural circumstances. In Brief 6 Ajai Nair describes leasing as an alternative to credit, which can help ease the provision of credit for investments in movable assets in rural areas. The brief describes the benefits of leasing to the client and the provider and identifies lessons on how to manage and support financial leasing in rural areas.

Finally, a key issue in financial service delivery is how to effectively increase repayment rates. In Brief 7 Yanyan Liu and Klaus Deininger discuss this issue in the context of self-help groups (SHGs) in India. Their analysis of the factors affecting repayment performance among low-income SHGs shows that effective application of rules pertaining to loan terms is more important than group characteristics in improving repayment performance.

Using modern communication technology

Recent advances in communication technology affect rural banking in two key ways. First, by facilitating electronic payment systems and branchless banking, this technology can significantly slash transaction costs for both service providers and consumers. Second, using portable smart technology to establish identification and monitor clients can significantly alleviate information asymmetries and help improve repayment rates.

In Brief 8 Susie Lonie describes how the cell phone–based payment service M-PESA now serves more than 9 million clients throughout Kenya, enabling them to remit money, make bill and loan payments, make cell phone–based payroll payments, and use banking services.

In Brief 9 Xavier Giné describes the results of an experiment to assess the impact of using biometric technology to monitor repayment performance of individuals in rural Malawi. This experiment showed that repayment rates increased by 40 percent for groups with a high default risk, and the benefits of improved repayment outweighed the cost of implementing the new technology.

Managing risk at the household and farm level

The management of risk is the key issue for financial institutions that finance agriculture, as well as for rural populations in general. In Brief 10 Mark D. Wenner analyzes various approaches to managing risk in financing agriculture. Index-based insurance schemes are one approach that has been implemented on a pilot basis in several countries. Such schemes use an easily observable index that is not subject to tampering. The index is correlated with the underlying risk and used to make decisions on insurance payouts, thus eliminating the cost of verification as well as incentives to misrepresent losses. In Brief 11 Jerry Skees and Benjamin Collier describe an ongoing pilot project in Peru that insures firms (such as microfinance institutions or firms in the value chain) serving smallholder households. The insurance pays out based on extreme El Niño events that create catastrophic flooding resulting in significant consequential losses and extra costs for a wide range of stakeholders in northern Peru.

Microinsurance has been developed as a risk management tool only recently. In addition to being more expensive to administer than savings and loan services, microinsurance schemes are plagued by more severe levels of adverse selection and moral hazard, which makes them challenging to provide on a sustainable, full-cost-recovery basis. Brief 12 by Martina Wiedmaier-Pfister and Brigitte Klein surveys key experiences in providing insurance in rural areas, including important issues related to regulating microinsurance.

Bundling financial and nonfinancial services

In addition to financial constraints, small farmers in developing countries also face market constraints in acquiring needed inputs (such as fertilizer, seeds, and extension services). Returns to financial services are thus highly conditional on access to other nonfinancial services. Brief 13 by Vijay Mahajan and K. Vasumathi describes how BASIX in India provides services such as soil testing and health monitoring of livestock, along with credit, to farmers in a way that maximizes returns to credit services.

Brief 14 by Jonathan Campaigne and Tom Rausch describes a similar approach used by the DrumNet project in Kenya. In contrast to BASIX, however, the DrumNet project uses information technology to link key actors along the supply chain to farmers.

The way forward

This set of briefs seeks to initiate discussions among stakeholders by disseminating information on a selection of innovative, on-the-ground initiatives designed to improve financial access for poor small farmers. All of these initiatives hold promise, but they also face challenges, and in the end some may not be suitable for a massive scale-up or for use in all country settings. Yet such initiatives demonstrate that it may be possible to eventually provide financing for agriculture on a sustainable basis at a reasonable cost.

Many of these initiatives are based on the premise that there is a supportive policy environment that allows innovation to flourish. The gravest risks to sustainable financing for agriculture often come not from inherent business risks or the inability of financial institutions to design profitable financial products for the rural population, but rather from misguided government interventions such as subsidized interest rates and lack of or non-enforcement of appropriate rules and regulations. Conversely, an enabling environment and legal framework, enforcement of regulations, and a supportive rural infrastructure would eventually lead to lower but sustainable interest rates by reducing transaction costs and risks and increasing competition. All this would contribute immensely to making sustainable access to finance a reality. ■

Renate Kloeppinger-Todd *(rkloeppingertodd@worldbank.org) is rural finance adviser in the Agriculture and Rural Development Department of the World Bank.* ***Manohar Sharma*** *(msharma5@worldbank.org) is senior poverty specialist in the Poverty Reduction and Economic Management Unit, East Asia Region of the World Bank.*

INTERNATIONAL FOOD POLICY RESEARCH INSTITUTE
Supported by the CGIAR
IFPRI® www.ifpri.org

THE WORLD BANK
www.worldbank.org

2020 VISION FOR FOOD, AGRICULTURE, AND THE ENVIRONMENT

Financial Literacy

MONIQUE COHEN

FOCUS 18 • BRIEF 2 • JULY 2010

The global financial crisis has intensified the problems of over-indebtedness, especially for the poor. In this context, the microfinance industry is giving more attention to building their customers' financial capabilities, designing products that respond to their needs and preferences, and ensuring their protection as consumers.

In a world where financial products and institutions are expanding rapidly, deciding which services to choose and how to use them is an increasing challenge. That challenge is especially great for customers who are poor and have limited experience in the formal financial sector. While money-management strategies can be innovative, the financial choices they make are defined by environments where informal financial practices are dominant and the consumer is often uncertain about commercial products and services. In increasingly complex and competitive financial markets, consumers with low levels of financial literacy lack the information and tools necessary to make informed decisions. Building financial capabilities can help people move from being overwhelmed by their financial options to being empowered by them.

Why is financial education important?

People at all income levels may have different resources and opportunities, but they still typically share common goals: They seek to put food on the table, educate their children, own a home, and plan for the future. To set aside even small amounts of money, low-income families need to be careful spenders as well as skilled money managers.

Financial education provides a foundation for managing money, which is an indispensable skill in a world where microfinance products and services are proliferating at the same time that overly aggressive financial services providers are ever ready to pressure the consumer. Building consumers' financial capabilities is about doing more with the little at hand, readying the unbanked (people without access to conventional banking services) to enter the formal financial system and enabling the underbanked (people with limited access to conventional banking services) to do more with the financial services at their disposal. It is also about improving the performance of financial services providers. Findings from a randomized impact evaluation found that Self-Employed Women's Association clients who attended financial literacy classes took out twice as many loans as women who did not (Pande, Field, and Jayachandran 2009).

How does financial education work?

Financial education is the process of building knowledge and skills to enable people to make more effective financial decisions while changing behaviors to build confidence in financial empowerment. The core of a financial-education agenda includes budgeting, saving, and managing debt. It also involves managing financial products such as insurance or remittances and making use of bank services.

Designing a financial-education program begins with a good understanding of the market. This means knowing the financial-literacy levels of the target population and the most effective delivery channels to reach them. Identifying the most appropriate "teachable moments" for financial education—for example, when someone first opens a bank account, starts a business, or makes a transition to technology-enabled banking—makes the education relevant and reinforces behavior changes since people have an opportunity to apply what they learn in the context of real life.

How can financial education be successfully delivered?

An important debate among practitioners is how financial education can be delivered most effectively. Channels range from public campaigns and mass media to face-to-face communication and personal counseling, from small-group seminars to classroom-style workshops. Innovative delivery channels also include cell phones and other electronic media.

Experience has shown that there is no best way to deliver financial education; it depends on the target group, objectives of a financial-literacy initiative, and available resources. Mass media—including television, street theater, call-in radio, or printed materials, such as posters and comics—is being used increasingly to expose poor and often illiterate people to key financial messages. Its primary impact is to spread awareness, whereas the purpose of face-to-face training and counseling is to provide participants with hands-on experience, particularly with banks, which they tend to distrust and fear. More fundamental changes in attitudes and behaviors require reinforced messaging over time.

Providers of financial education have differing interests, which translates into a diversity of delivery approaches. Central bankers or regulators who wish to protect consumers from fraud and abuse tend to give priority to public campaigns focused on consumers' rights and responsibilities. Financial institutions that aim to increase adoption and use of their products and services may choose to integrate financial-education messages into their marketing agendas. Community-based organizations wishing to promote livelihoods and asset building for the poor may integrate financial education into a range of activities, including extension services, health education, business-development training, or mentoring. Consumer-protection organizations may embrace financial education as part of social-marketing campaigns, community-based training, or one-on-one counseling at debt advisory centers.

The choice of delivery systems is very much a question of resources. While tangible, direct training is expensive on a large scale, bundling delivery channels—for example, combining radio with some direct training offers—can help strike a balance between achieving both broad and focused impacts.

Outcomes and impact

Controversy surrounds the issues of what and how to measure the outcomes and impact of financial education. (See Figure 1 in Appendix A for one such approach.)

Currently, quantitative evidence of the positive outcomes and impacts of financial education is limited. This contrasts with affirmative anecdotal evidence from learners. Meanwhile, research shows that financially literate clients make better financial decisions and maintain a better overall financial well-being (Cole and Fernando 2008).

Recent research linking financial education to behavior changes among low-income microfinance clients in Bolivia and Sri Lanka provides insights into these contradictory impact-related observations. Two years after receiving financial education, clients increased their knowledge of loan products and debt capacity. Positive changes in savings behaviors included reducing expenses as well as recognizing the value of saving three times the amount of monthly income for emergency purposes. Those given budgeting training identified the primary function and different parts of a budget and were able to work within their own budgets. However, putting debt-management and savings behaviors into practice during the food and financial crises that affected these countries was a challenge. The new savings behaviors translated into reduced vulnerability (Gray et al. 2010).

To assess the outcomes of financial education, researchers must look beyond indicators of behavior change. They must recognize that financial behaviors are influenced by the context in which people live—both inside and outside the household—and thus are ever changing. According to Gray et al. (2010), the five elements of effective financial education are

- quality and frequency of education,
- relevance of the education to the target population,
- opportunity to use the education,
- context in which people can exercise their new financial behaviors, and
- appropriateness of the financial products offered.

Developing a financial education agenda

Since 2002, Microfinance Opportunities (MFO) has sought to put financial education on the agenda of microfinance institutions and other development organizations seeking to improve the financial lives of the poor. To this end, MFO partnered with Freedom from Hunger to develop a global financial-education curriculum that is targeted at those just above and below the poverty line in developing countries. Developed in partnership with nearly 20 microfinance service providers, the curriculum currently addresses ten themes. The core topics are budgeting, saving, managing debt, negotiating financial transactions, and using bank services. Each theme includes (1) a content note that provides a topic overview, (2) a trainer's guide with step-by-step instructions for conducting each learning session, and (3) a "training of trainers" manual to prepare financial-education trainers.

The trainer's guide, the cornerstone of the curriculum, has proven itself a valued reference tool. The curriculum itself is flexible and readily adaptable to longer or shorter learning activities, different contexts, and target populations, including people who are illiterate (Nelson and Wambugu 2008). The base curriculum has been expanded to encompass a number of specialized modules focused on particular financial products (including insurance, remittances, and housing loans), specific target groups (for example, adolescent girls) and consumer protection. By leveraging partnerships and using scalable delivery channels, Microfinance Opportunities is achieving significant levels of outreach; in less than three years, more than 500,000 consumers have received direct training. This figure does not include the enormous outreach achieved through the viral spread and adaptations of the MFO curriculum using mass media. For example, Makutano Junction, a televised series in East Africa, has incorporated the key messages into several episodes. Its viewers are in the millions.

Where do we go from here?

Financial education is beginning to get the visibility and interest it deserves. Attention is moving beyond the implementation of small-scale initiatives to the development of national financial-literacy strategies that straddle financial and social policies. Integration of financial education into cash-transfer programs and branchless banking are other emerging programmatic areas. Measuring how and when financial education translates into financial-behavior change remains difficult, but, among learners, it is valued and has emerged as a key—although often overlooked—component of economic empowerment.

Some see the challenge of scalability as an obstacle to a meaningful impact of financial education. It need not depend on stakeholders' objectives for financial education. Everyone can benefit from financial education: the banked, unbanked, or underbanked. Technology offers just one avenue to send key messages to large numbers of people; its spread therefore must not be restricted to users of formal financial services. Building financial capabilities among the low-income population is a win-win situation for the financial sector because it creates better-informed consumers. Financial education need not be a stand-alone activity. It is very effective when combined with other development interventions aimed at reducing vulnerability and food insecurity and expanding opportunities for the poor. ■

See more information at www.microfinanceopportunities.org.

For further reading: S. Cole and N. Fernando, "Assessing the Importance of Financial Literacy," ***ADB Finance for the Poor*** **Vol. 9 (No. 3): 2008; B. Gray, J. Sebstad, M. Cohen, and K. Stack "Can Financial Education Change Behavior?: Lessons from Bolivia and Sri Lanka," Working Paper 4 (Microfinance Opportunities, Washington, D.C.: 2010); C. Nelson and A. Wambugu,** ***Financial Education in Kenya: Scoping Exercise Report*** **(Financial Sector Deepening Kenya, Nairobi, Kenya: 2008); A. Klincic, "Case Study of Opportunity Bank of Malawi" (Microfinance Opportunities, Washington, D.C.: Forthcoming); R. Pande, E. Field, and S. Jayachandran,** ***Business Training and MFI Client Behavior: Findings from a Randomised Impact Evaluation in Ahmedabad, Gujarat*** **(Institute for Financial Management and Research (IFMR) Centre for Micro Finance, Chennai, India: 2009).**

Monique Cohen *(moniquec@mfopps.org) is the president of Microfinance Opportunities.*

2020 VISION FOR FOOD, AGRICULTURE, AND THE ENVIRONMENT

Community-Based Financial Organizations: Access to Finance for the Poorest

ANNE RITCHIE

FOCUS 18 • BRIEF 3 • JULY 2010

Community-based financial organizations (CBFOs) are user-owned and -operated groups that provide mainly saving and lending services but may also offer other financial services such as insurance. These independent organizations are based in local communities, with local governance and management. CBFOs range in size. They can take the form of informal and unregistered groups of five to seven people, usually women, who meet weekly to save small amounts of money that they then lend to each other and possibly to other members of the community. They also include larger, slightly more formal groups of up to 40 people who have written by-laws, and they include small financial cooperatives. CBFOs flourish among people who have poor access to banks and nonbank financial institutions such as microfinance institutions (MFIs).

Market niche

The market niche served by CBFOs is the unbanked poor. In many countries, locally organized CBFOs, such as rotating savings and credit associations (ROSCAs), have served as financial intermediaries for their communities for generations. ROSCA members save a predetermined amount of money regularly. In each period, one member of the ROSCA receives the funds collected. ROSCAs thus allow people to accumulate, through small regular savings, a large lump sum that is available for investments, such as creation or expansion of small businesses, children's education, and home improvement. The main drawback of ROSCAs is that the money may not be available when needed because only one member collects the funds at one time.

Although MFIs formed over the past four decades have done a great deal to make financial services available to the unbanked poor, they have not, for the most part, been able to reach the poorest people, especially those who live in remote rural areas. The poorest are able to save and borrow only very small amounts of money, making it too costly for banks and MFIs to serve them. In remote rural areas with widely dispersed populations, banks and MFIs often cannot cover the costs of an extension agent or a branch office, even if they use modern technologies to reduce costs or group people together to achieve economies of scale. Thus, MFIs have been successful in broadening the number of people served but less successful in reaching the poorest.

Successful models

Experience has shown that successful CBFO models must incorporate a number of basic principles: social cohesion of group members, a focus on building up savings to fund loans rather than relying primarily on external sources of funds, and an organizational structure that enables governance and management by people who are often poorly educated and have little or no experience with financial management beyond managing their own households and economic activities. Two models in particular appear to work well on a large scale and have good prospects for long-term sustainability.

One model is the village savings and loan association (VSLA) model. Started in Niger by CARE International in 1991, the VSLA adopted lessons from the efforts of poor local women to save in this large, poor, sparsely populated country. Since then, CARE and other nonprofit development agencies have spread the model to 39 countries, the vast majority in Africa. VSLA groups, consisting of between 10 and 30 members, have simple rules that govern their savings and lending activities. Each member saves on a regular basis, and this money is then lent out at an interest rate and on loan terms decided by the group. Loans may be made to both members and nonmembers. Indeed, many members save but do not borrow and earn a good return on their investment through the interest charged to borrowers. At the end of a given period, usually a year, the savings and the interest the VSLA earned are distributed to the members, and a new cycle begins. The distribution feature of this model keeps the amounts of money that the members must manage at a level commensurate with their financial literacy. It also enables all members to receive a lump sum on the same date, often one that coincides with most members' need for funds, such as an annual festival, the start of the planting season, or the date that school fees must be paid. VSLAs do not generally link with banks or MFIs because experience has shown that members' savings are generally sufficient to meet their credit needs, and injection of external loan funds has caused many groups to fail.

The self-help group (SHG) model, begun in India several decades ago, has become the dominant microfinance model in that country, especially for the rural poor. SHGs usually have between 10 and 20 members who save regularly and lend the money out to members only. The funds saved are not distributed back to members, but, rather, grow over time. SHGs in India often receive small amounts of seed capital from government or donors. They usually have an explicit goal of bank linkage, which has been facilitated by the high density of banks in rural areas and by a government policy stipulating that banks' portfolios must include rural loans. Many SHGs belong to federations that provide them with access to external capital, technical assistance in areas such as accounting, and greater bargaining power with government and banks. As of 2007, India had approximately 69,000 SHG federations.

The principal differences between the models are the following:

- VSLAs are self-contained at the village level, whereas SHGs link with banks and form federations with other villages.
- VSLAs distribute all savings and earnings back to members at the end of the year, whereas SHGs add new savings to existing savings with no automatic distribution mechanism. This difference makes VSLAs easier for nonliterate people to manage but allows SHGs to accumulate more capital for lending.

Matching CBFO models with communities' needs

The design of a CBFO program should be responsive to prevailing local conditions. A number of factors should be taken into consideration, including the demand for financial services and the proximity of banks and MFIs. In poor rural areas with weak local economies dominated by subsistence farming and few new business opportunities, VSLAs can provide an effective way for households to manage their financial resources. Savers are able to earn a return on their investment by making their capital available to those with viable businesses. If banks and MFIs are distant, as in the rural areas of many African countries, attempting to foster bank linkage may be more expensive than is warranted by the limited demand for loans.

In areas with more vibrant economies and greater population density, the bank linkage and federation aspects of the SHG model enable groups to draw on external funds for the growth of members' businesses. Federations can help SHGs with financial management and may also offer training aimed at strengthening the SHGs. However, because both the bank linkage and federation aspects of the SHG model add significant levels of complexity, external support from a technical-assistance provider may be required for a long period of time.

Both VSLAs and SHGs are initially formed and nurtured by trained extension agents. Experience with both models has shown that once the model has been established in a particular area, setting up new groups can be less expensive because members of existing groups can spread the model to other communities through informal linkages between communities or through the formation of associations of trainers who are themselves group members.

Leveraging finance and partnerships with mainstream financial institutions

The question of external financing has generated great debate. Many CBFOs have failed following the infusion of donor or government funds into fragile young organizations lacking the skills to manage this money. External credit may also draw into the membership people whose main objective is to obtain a slice of donor largesse rather than to contribute to the slow but steady buildup of the group through its own efforts. Yet it is precisely these efforts that are needed to build effective governance and management.

Nevertheless, partnerships between CBFOs and mainstream financial institutions can be beneficial, especially if implemented incrementally. In the simplest form of partnership, CBFOs may bank their excess savings and earn interest on these savings. As the relationship develops, the bank or MFI is able to assess the capacity of the CBFO to manage its own funds. In a World Bank–supported project in Sri Lanka, rural banks have been eager to develop relationships with CBFOs, which provide banks with easy access to a large number of rural customers. In some cases, the banks have sent their representatives to the villages to open the bank accounts. Such confidence-building measures can, over time, lead to a willingness on the part of the bank or MFI to extend credit to either the CBFO (for on-lending to its members) or to individual members who have viable business plans.

The way forward

The ability of CBFOs to govern themselves effectively and to manage their operations so that savings are secure and loans are repaid is paramount for their long-term sustainability. Donors and government can add value by funding programs that train local people to develop viable groups and by providing technical assistance for the development of simple governance, operational, and accounting systems that can be implemented locally.

Donors and governments should also fund program evaluations, using performance criteria that allow comparison across programs and models. The single most important performance indicator is repayment performance—that is, the ability of CBFOs to get borrowers to repay their loans in a timely way. Nonrepayment of loans is the greatest threat to the financial sustainability of any financial organization, including CBFOs. This threat is increased by the tendency of donors and governments to provide CBFOs with large loan funds that are beyond their capacity to manage effectively. Significant amounts of external funding—beyond small seed funds that help groups get started—should be linked to their performance in managing the group's own funds. This careful approach will enable CBFOs to develop a strong foundation that enhances their prospects for long-term sustainability. ■

For further reading: J. Murray and R. Rosenberg, Community-Managed Loan Funds: Which Ones Work? Focus Note No. 36 (Washington, DC: Consultative Group to Assist the Poor, 2006), www.cgap.org/p/site/c/template.rc/1.9.2577/; A. Ritchie, Community-Based Financial Organizations: A Solution to Access in Remote Rural Areas? Agriculture and Rural Development Discussion Paper 34 (Washington, DC: World Bank, 2007); APMAS, www.apmas.org/; Gemi Diriya Foundation, www.gamaneguma.lk/sub_link_view.php?doc=19; VSL Associates, www.vsla.net/.

***Anne Ritchie** (AnneFRitchie@aol.com) is a World Bank microfinance consultant.*

INTERNATIONAL FOOD POLICY RESEARCH INSTITUTE
Supported by the CGIAR
IFPRI® **www.ifpri.org**

THE WORLD BANK
www.worldbank.org

Rural Banking in Africa: The Rabobank Approach

GERARD VAN EMPEL

FOCUS 18 • BRIEF 4 • JULY 2010

Many people in the vast rural areas of Africa lack access to financial services, and most commercial banks are not interested in moving into these areas due to their low income levels, lack of scale economies, and poor infrastructure. Also, few banks actually understand the most common economic activity in rural areas: agriculture.

Consequently, the absence of financial institutions in rural Africa has often enticed governments to step in, particularly with state-dominated banks focused on agriculture. Many of these initiatives have failed, however, because they were too bureaucratic, too policy oriented, too concentrated on risk to only one segment of the population, or too weak in customer focus. In addition, clients considered these government-sponsored institutions to be instruments that provided grants; hence, the banks suffered from poor loan-recovery rates.

While microfinance institutions have made some inroads into rural Africa with the financial backing of international nongovernmental organizations and other sponsors, their sustainability is questionable. They tend to lack banking licenses and therefore have a very limited product range, and they cannot afford modern technology-based distribution systems.

Key gaps in rural banking in Africa

One of the most prominent gaps in developing banking services for rural Africa is poor infrastructure—for example, bad roads, erratic electricity provision, and lack of communications systems—which impedes effective outreach to customers.

The legal environment in these rural areas is also suspect. Insecure property rights—especially land titles in rural areas—limit any bank's collateral options; combined with poor contract-enforcement opportunities, this takes away a bank's incentive to provide credit, especially for long-term loans. Proper land registration and enforceable mortgage systems are important issues for rural development.

The inefficiency of markets is also a barrier to developing rural financial services. Agricultural value chains are often poorly organized, lacking in transparent pricing, and fragmented in primary production—all of which results in high transaction costs. In many cases, the banking environment is distorted by stakeholders—including donors, governments, and development banks—who do not always regard agriculture as an economic activity, but rather as a social problem. These stakeholders provide subsidized funding to farmers or cooperatives, which means private banks often lack a level playing field. Poor financial literacy rates, especially among small farmers, and a limited understanding of banking requirements also pose a problem.

Rabo Development

In an effort to serve the financial needs of emerging markets and developing countries, Rabo Development (RD) was created with a mission similar to that of its parent organization, Rabobank, which was created by farmers in the Netherlands more than one hundred years ago. In order to achieve its mission of providing access to financial services to those in rural areas of developing countries, RD participates in financial institutions and provides management services and technical assistance. It has made investments in Tanzania, Zambia, Mozambique, Rwanda, Paraguay, Brazil, and China. RD also works with cooperative "enterprises" and financial institutions that want to increase their own access to financial services. While RD focuses on investments, Rabo International Advisory Services (RIAS) provides technical assistance. RIAS has a 20-year history in consultancy services mainly to financial institutions and cooperatives in emerging markets.

Rabo Development has made investments in some existing financial institutions with the objective to transform these organizations into leading banks with a rural orientation. Financial participation is limited to minority stakes (variation between 10 and 45 percent); the majority of shares are locally owned, thereby retaining the status of a local bank. This allows customers to better identify themselves with the institution and appeals to national pride.

Financial investment demands a shared vision of the future development of an institution, so stakeholders should agree on a mission and business plan, which are likely to entail servicing new customer segments, including rural clients (mainly farmers), and developing a broader product base. In order to assist implementation, Rabo Development not only provides capital but also management services, technical assistance, and representation on the board of directors. At present, Rabo Development has four investments in banks in Africa, which have a total of more than 3 million customers collectively.

There is no standard recipe for exactly how capacity should be built because it very much depends on each individual institution's stage of development and the country concerned. In any case, however, special attention should be given to getting the product distribution strategy of the institution right. To distinguish customer segments and develop their value propositions, it is pivotal to organize an efficient outreach, using both physical channels (such as branches) and virtual channels (such as ATMs, mobile banking, and Internet banking).

Lessons learned

These lessons have been learned by Rabo Development's work in developing countries and emerging markets.

- Banks with a rural orientation still need a strong urban presence, as most banking assets are concentrated in urban areas. New distribution concepts, such as mobile banking and products, are also normally piloted in urban areas.
- Banks need to service all client segments with the appropriate mix of products (including microloans) in order to effectively use the branch network and establish a well-balanced portfolio, thereby reducing the concentration of risks.
- Financial institutions that want to practice rural finance need to be committed to this segment and need specialized knowledge-based departments, including agriculture and small

and medium enterprises (SMEs), in order to be effective to these target groups. Political pressure has forced institutions in some countries to provide rural financial services, but, with no commitment, results are poor.

- Sufficient scale and market share is essential for banks with a rural orientation.
- Client linkage to corporate governance—for example, rural and urban client shareholders or client panels—can positively influence performance and safeguard the rural mission of an institution.
- Clear client segmentation linked to products and distribution channels is essential to effectively delivering products in the rural environment.
- Direct product distribution channels—including mobile-phone banking, ATMs, and electronic point-of-sale devices—are becoming increasingly important for rural finance delivery. A modern and up-to-date IT system is crucial to these services.
- Rural banks in Africa need to focus on both sides of the balance sheet (that is, offer an appropriate mix of savings and lending products). Due to a lack of well-operating markets, they need to be largely self-financing.
- Policy instruments based on risk or cost sharing can be effective but need to be based on clear client segmentation. In cases of sufficient payment capacity, they can be used to overcome the lack of enforceable collateral or to make the credit itself more enforceable. Clients with insufficient payment capacity can only be assisted through income-support mechanisms, meaning grants not loans.

The Rabobank approach to financing agriculture

Among the most important lessons Rabo Development has learned about building sustainable agrifinance in Africa is that segmentation of farmers is essential. In general, a small group of commercial farmers is responsible for a large part of a country's agriculture production and exports, and often these farmers are the only ones with access to financial services. A large group of subsistence farmers who lack sufficient repayment capacity for bank loans resides at the bottom of the pyramid. The group between the subsistence farmers and the commercial farmers consists of both farmers of small cash-crops (for example, coffee, cotton, or cocoa) with a low annual marketable surplus and so-called "emergent" farmers. The latter group has the potential to grow into commercial farmers but lacks both the financing and farm-management expertise.

Financing small cash-crop farmers is only feasible through a supply-chain approach. These smallholders should be financed indirectly via contract farming with better-rated "off-takers" (or processors). Under such schemes, the farmer commits to supply 100 percent of a particular crop to the off-taker, and the off-taker commits to buy 100 percent of the farmer's product but pays that money directly to the bank, thereby allowing a direct repayment. Under these structures, the repayment risk to the individual farmers is converted into performance risk to both the farmer and the off-taker. In many cases, cooperatives can play a facilitating role by being the counterpart of the off-taker and the borrower of the loan. A cross-liability system whereby the members guarantee one another's loans could provide extra comfort to the bank. Also, systems involving warehouse receipts can provide additional financing to this target group; they have been used effectively by the banks that Rabo Development has invested in.

Emergent farmers justify an individual approach since they have the potential to develop into commercial or professional farmers with corresponding growth of financial services. Strict criteria need to be established regarding minimum size, sufficient entrepreneurial spirit, basic understanding of business planning, and farm-management skills. With a combination of financial services and technical support, these farmers stand a fair chance of success.

Emergent farmers can be financed under the existing retail structure of a particular bank, but the local branches involved would need to hire and train agrifinance specialists who understand farming and have the ability to appreciate the particular risks associated with it (including, among others, climatic, disease, and price risks). It is essential to form alliances with other stakeholders in the value chain who also have an interest in developing and investing in the farming sector (for example, farmers' organizations, commodity exchanges, agri-input providers, and off-takers).

The main obstacles to financing agriculture are unpredictable or erratic government behavior and interference in the agricultural sector. This is especially the case in cash crops like coffee, which are often important sources of hard currency, and in grains, of which African countries are often net importers. In several coffee-exporting countries, the coffee export is not free but rather regulated through auctions with only a limited number of private exporters licensed. In grains, prices are often regulated by the government to safeguard food security. This comes often at the expense of local farmers who are struggling to break even, and it is aggravated by relatively high transaction costs and the weak market position of African farmers.

Conclusion

The Rabobank approach is strongly focused on the value chain, as ultimately the farmer—who runs the price risk, to a large extent—will only be able to get a fair price when the whole chain operates effectively. The success of agricultural development depends on the creation of a large group of professional local farmers producing high volumes of marketable output at a consistent quality. This will have a positive effect on reducing the transaction costs throughout the whole value chain. It is also imperative that all those involved share a common vision on development and contribute in effective, constructive, and committed ways. ■

See more information at www.rabobank.com.

Gerard van Empel (g.j.j.m.empel@rn.rabobank.nl) is general manager of Rabo International Advisory Services and director of Rabo Development.

INTERNATIONAL FOOD POLICY RESEARCH INSTITUTE
Supported by the CGIAR
IFPRI® www.ifpri.org

THE WORLD BANK
www.worldbank.org

Rural Banking: The Case of Rural and Community Banks in Ghana

AJAI NAIR AND AZEB FISSHA

FOCUS 18 • BRIEF 5 • JULY 2010

Before the late 1970s, rural dwellers in Ghana had almost no access to institutional credit for farm and nonfarm activities, and in many rural communities, secure, safe, and convenient savings and payment facilities hardly existed. In response to this situation, the Government of Ghana took several measures to increase access to credit in rural areas, including facilitating the establishment of rural and community banks (RCBs). This brief discusses the history of RCBs, their business model, their services, and their financial performance. It then draws some lessons relevant for others involved in or planning similar initiatives.

As a network, RCBs are the largest providers of formal financial services in Ghana's rural areas. By the end of 2008 Ghana had 127 RCBs with a total 584 service outlets, representing about half of the total banking outlets in the country. The RCB network reaches about 2.8 million depositors and 680,000 borrowers. Although the service delivery performance of the RCB network has been strong, its financial performance has been mixed. The profitability and net worth of the network have grown, but the financial performance of some members has been poor, and a small number are insolvent.

The creation and evolution of the RCBs

The first RCB was established in a farming community in the Central region of Ghana in 1976. Several others were established in rapid succession, and by 1984 the number of RCBs reached 106. By the early 1980s, however, the financial performance of many RCBs started to decline for several reasons, including a 1983 drought, weak governing ability, conflicts within boards of directors, and ineffective management in many RCBs.

The Bank of Ghana, the Ghanaian central bank, undertook several reforms to curb the deteriorating situation. Exposure to risky sectors (mainly agriculture) was limited, distressed banks were closed, supervision was strengthened, and RCB managers and boards of directors were offered training. Between 1989 and 1994 the Government of Ghana, with the support of the World Bank, also implemented the Rural Finance Project, aimed at providing targeted support to the RCBs. The project contributed to an improvement in RCB performance.

Nevertheless, several RCBs remained weak, and in 1998, the Bank of Ghana liquidated 23 RCBs. The Government of Ghana, with the support of the World Bank and other donors, implemented a follow-up project—the Rural Financial Services Project—between 2001 and 2007 to help further strengthen the RCBs. This project provided extensive training to RCBs and supported the establishment and strengthening of the Association of Rural Banks (ARB) Apex Bank, as a bank to the RCBs. (The Association of Rural Banks had been established in the early 1980s as a networking forum for RCBs and later started providing training to member RCBs.)

Business model

Small asset base. RCBs are relatively small financial institutions with average share capital of GHc 136,526 (US$105,263), average deposits of GHc 2.3 million (US$1.77 million), and average assets of GHc 3.8 million (US$2.4 million), although values of the three indicators vary significantly among RCBs.

Community ownership and governance. RCBs are fully owned by shareholders who are residents of communities in which they operate. Each rural or community bank has a board of directors that is responsible for its strategic governance. Boards are elected by owners/shareholders during annual general meetings. Election criteria are normally based on reputation in the community and professional expertise, but experience in banking is extremely limited.

Professional management and staff. The core management staff of a typical RCB is composed of a chief executive officer who is in charge of the daily management of the bank; an internal auditor, responsible for internal control measures; a finance officer; and credit and project officers. Many of the personnel are recruited from local communities.

Strategic alliance. Since 2002 the ARB Apex Bank has provided specialized services essential to improving the quality and scope of products offered by RCBs, and it performs important supervisory functions delegated by the Bank of Ghana. Among the main services offered by the Apex Bank are check clearing, specie supply, treasury management, loan fund mobilization, and domestic and international money transfers. The Apex Bank provides most of these services on a fee basis.

Legal and regulatory framework. RCBs are incorporated as limited liability companies and licensed by the Bank of Ghana within the framework of the Banking Act. The minimum level of capital required by RCBs is GHc 150,000 (US$116,135). RCBs whose capital falls below this minimum are not allowed to pay dividends or open new branches or agencies until they attain the minimum level of capitalization.

Products and services

Savings. RCB savings products include savings accounts, current accounts, susu deposits (small savings collected daily from clients by individual collectors going door to door), and fixed or time deposits. In a sample of 12 RCBs, regular savings deposits account for about 58 percent of the total number of clients and 57 percent of the total deposit balance. These accounts are small in size and short term. Susu is the second-largest account type, representing 21 percent of total clients, but its share of total deposits is only 11 percent because of the small size of each account. Fixed and special deposits that offer higher interest rates with long-term deposit contracts represent only about 1 percent of total clients.

Credit. The credit products offered by RCBs include microfinance loans, personal loans, salary loans, susu loans, and overdraft facilities. In a sample of 12 RCBs, salary loans amount to 33 percent of total advances, followed by personal loans (24 percent) and microfinance (20 percent). In terms of number of borrowers, microfinance accounts for 31 percent of total borrowers followed by personal loans (26 percent) and salary loans (22 percent). RCB loans are used for agriculture, cottage industries, and trading.

Money transfers and payments. RCBs participate in local and international money transfers, and government agencies use the RCB service outlets for salary and pension deposits. Clearing of checks for cocoa purchases is also an important service provided under the payment category.

Performance

Steadily increasing outreach and service delivery. Between 2000 and 2008 the number of depositors in RCBs grew at an average annual rate of 14 percent, and the number of borrowers grew at an average annual rate of 27 percent. The RCB network reaches about 2.8 million depositors and 680,000 borrowers, making RCBs the largest group of licensed financial service providers in rural areas. Clients of RCBs consist mostly of farmers, government employees, and small and micro-entrepreneurs.

Mixed financial performance. The profitability and net worth of the RCB network steadily increased from 2000 to 2008. Network-wide capital is well above the minimum 10 percent required by the Bank of Ghana. In 2008, however, seven RCBs were insolvent, and the continued operation of poorly performing RCBs is a key issue facing the network. The relatively high ratio of nonperforming loans is another major factor affecting financial performance. In the sample RCBs, for example, the proportion of the loan portfolio that was in default for more than 30 days was 16 percent, compared with 3 percent for banks in their global peer group.

Lessons on rural banking

The case of rural banking in Ghana points to the following lessons:

- Although community-based financial institutions such as the RCBs can play a key role in increasing access to financial services in rural areas, their small size can also make it challenging for them to become financially sustainable and compete with other financial institutions that enter the rural financial market. To be successful, they should be able to respond dynamically to changes in the business environment. These responses may include building linkages, being open to mergers, and bringing in external investors, if necessary.
- Small local financial institutions often cannot easily procure needed technical support (such as training and specialized technical assistance for product development and setting up of operational systems) from the market. Hence, initiatives to build local financial institutions must support the creation of strategic alliances that can either provide such services or facilitate their cost-effective provision. Apex institutions can play a crucial role in providing technical and financial services to small financial institutions.
- Apex institutions may find it difficult to achieve financial sustainability by providing services to members alone. Such institutions may have to also provide services to the public, including general commercial banking services. Care must be taken, however, to ensure that the business model adopted does not compromise the original mission—in this case, to increase sustainable provision of financial services in rural areas.
- The regulator needs to have the necessary skills, political autonomy, and financial resources to effectively regulate and supervise a large number of small financial institutions that are geographically dispersed. Often the central bank does not have the skills to undertake this task directly, and alternative models of supervision may have to be adopted. Even in the best circumstances, however, a certain number of institutions will fail, and the regulatory system needs to have the capacity to respond quickly to protect depositors and to prevent failure from lowering confidence in other institutions. Donor funding cannot sustain a supervisory regime in the long run, and recovery of all supervision costs through fees from the supervised institutions may not be a feasible option. Under these circumstances, adequate government funding for supervision would be critical for ensuring sustainable service delivery. ■

For further reading: A. Nair and A. Fissha, "Rural Banking: The Case of the Rural and Community Banks in Ghana," Agriculture and Rural Development Discussion Paper No. 48 (Washington, DC: World Bank, 2010).

__Ajai Nair__ (anair@worldbank.org) is program coordinator of the Agriculture Finance Support Facility in the Agriculture and Rural Development Department of the World Bank. __Azeb Fissha__ (afissha@worldbank.org) is a consultant with the Agriculture and Rural Development Department of the World Bank.

INTERNATIONAL FOOD POLICY RESEARCH INSTITUTE
Supported by the CGIAR
IFPRI® www.ifpri.org

THE WORLD BANK
www.worldbank.org

Rural Leasing: An Alternative to Loans in Financing Income-Producing Assets

AJAI NAIR

FOCUS 18 • BRIEF 6 • JULY 2010

Credit for investments that pay back in the medium to long term (three to five years or longer) is in short supply in rural areas. Credit unions and microfinance institutions (MFIs), which generally have better outreach than commercial banks in rural areas, typically provide only short-term credit. Credit available from informal sources (such as moneylenders, family, and friends) is usually both short term and too costly for investment financing. For rural enterprises seeking to acquire equipment—a typical investment need—to modernize production and thereby increase productivity, one solution may be financial leasing.

Leasing offers several advantages. For traditional credit, farmers and rural enterprises are particularly constrained by a lack of assets that can be used as collateral. Leasing overcomes this constraint because it requires no collateral or less collateral than typically required by loans. Because leases also often require lower down payments than the equity required for loans, they are more affordable for rural enterprises that have limited funds and little access to borrowed funds.

From the lessor's perspective, not having to obtain collateral is particularly advantageous in a rural context. Although the difficulties involved in creating, perfecting, and enforcing security are applicable in both urban and rural contexts in most developing countries, they are more severe in rural areas where enterprises are less likely to hold titles to their assets, asset registries are less likely to be functional, and judicial processes are likely to be slower. Lessors are also likely to benefit from not being restricted by interest rate ceilings and sector-specific credit allocations—factors that have traditionally constrained rural lenders. Boxes 1 and 2 explain key features of a leasing contract, and Figure 1 shows a typical tripartite financial lease transaction involving an equipment supplier, a lessor, and lessee.

Box 1—What is a financial lease?

Leasing is a contract between two parties, where the party that owns an asset (the lessor) lets the other party (the lessee) use the asset for a predetermined time in exchange for periodic payments. Leasing separates use of an asset from ownership of that asset. There are two main categories of leasing: financial leases and operating leases.

In a financial lease, lease payments amortize the price of the asset. At the end of the lease period, the lessee can purchase the asset for a token price. The lessee is responsible for maintenance and risk of obsolescence of the asset. Because of the option to purchase the asset and the risks transferred to the lessee, a financial lease is a close substitute for a loan. Nearly all rural leases are financial leases.

In contrast, operating leases do not include the option to purchase the asset. Maintenance costs and risk of obsolescence are borne by the lessor, and leases are cancelable.

Box 2—Key features of a financial lease contract

- Security: The primary security is the leased equipment. In some cases a small amount of cash or other asset owned by the lessee may be taken as additional security.
- Insurance: The lessor insures leased assets with commercial insurance and includes the cost in the lease price.
- Lease term: Lease terms range from two to five years.
- Lease cost: It includes cost of insurance, operating cost, loss provision, and profit.
- Lease payment schedule: The payment schedule can be monthly, quarterly, half-yearly, or annual.
- Option to purchase: On completion of the lease payments, lessees have the option to purchase the leased assets at a certain percentage of the lease cost.

Figure 1—Financial lease transaction

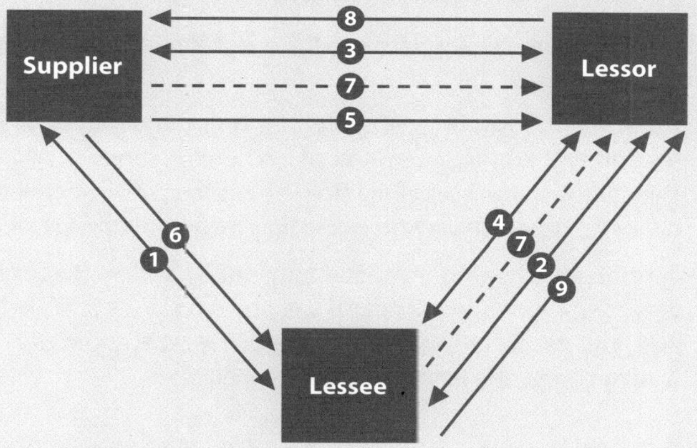

1. Initial negotiations about model, specification, price, discounts, warranty, delivery, etc. At this time the method of payment for the asset may not have been discussed.
2. Request for a leasing quotation (the supplier may also provide quotations on behalf of lessors).
3. Purchase contract agreement signed between lessor and supplier based on information supplied by the lessee to include those issues in (1) and also payment terms.
4. Lease contract signed and downpayment paid by lessee.
5. Invoice created by supplier giving title in asset to lessor (assuming full payment received by supplier).
6. Asset delivered to lessee.
7. Delivery and acceptance notice (protocol) signed by supplier and lessee.
8. Supplier's invoice paid by lessor.
9. Regular lease repayments paid.

Source: IFC (International Finance Corporation), *Leasing in Development: Guidelines for Emerging Economies* (Washington, DC, 2009).

Rural leasing initiatives

A 2006 World Bank case study of three profitable providers of leasing in rural areas showed that in all three cases the rural portfolios were as profitable as their urban portfolios. Arrendadora John Deere, the largest provider of farm machinery leases in Mexico, had nearly US$63 million in farm equipment leases. DFCU Leasing, the largest provider of leases in Uganda, had a US$5 million lease portfolio in rural areas. Network Leasing Corporation Limited, a leading micro-leasing provider in Pakistan, had a lease portfolio of more than US$2.4 million in rural areas. Low lease losses, strong client demand for asset financing, and a favorable legal and policy environment made rural leasing a profitable business for these companies. For clients, access to finance at a reasonable cost, low or no collateral requirements, quick processing, and easy access to the provider appear to be significant benefits.

Drawing on the experiences of the providers studied, the World Bank study identified the following lessons on managing financial leasing in rural areas.

- **Rural leasing is a means to acquire productive assets.** All rural leases provided by the three leasing companies are financial leases and were used to finance the acquisition of assets (in contrast to renting of assets).
- **Rural enterprises of different sizes benefit from leasing, but a provider may not be able to equally serve all enterprises.** Providers are limited because of differences in the skills and capacities required to effectively serve enterprises of varying sizes.
- **Nonfarm enterprises account for a significant proportion of rural leases.**
- **Rural leasing can be profitable, but jump-starting rural leasing will require government and donor support.** All three firms studied benefited from access to government or donor funds, particularly in expanding their rural operations.
- **A rural-only leasing company may not be viable.** Because leasing is a specialized financial activity, economies of scale, cost, and risk factors may require that, in most economies, leasing companies have larger urban operations.

The challenge: Supporting increased availability of leasing in rural areas

Leasing is a viable tool to finance rural assets. The nature and capacity of existing financial institutions, the level of potential demand for investment finance in rural areas, and the level of development of the leasing industry should determine the mechanisms for supporting increased access to leasing for rural enterprises. Policy-level support will be required in countries that do not have a clear legal and regulatory framework for leasing. Such support must be sectorwide and not restricted to rural leasing.

A good legal framework for leasing includes (1) clear definitions of a lease contract, leased assets, and responsibilities and rights of the parties to a lease contract; (2) clarity in allocating responsibility for liability for third-party losses arising out of the operation of leased assets; (3) stipulation of the priority of a lessor's claim over a leased asset; and (4) a framework for easy and fast repossession of leased assets. The use of internationally accepted accounting standards and an unbiased tax code enhance the development of the leasing sector. The existence of a well-functioning asset registry, the availability of insurance and maintenance services for equipment at a reasonable cost, and the existence of a good market for used assets are also necessary for the development of the financial leasing industry.

Targeted institutional support may also be needed to help develop the rural leasing sector. As shown in Boxes 1 and 2 and Figure 1, financial leasing is a relatively complex transaction. To successfully undertake financial leasing operations, organizations need not only well-trained staff, but also high-quality lease origination processes, accounting and internal control systems, and overall portfolio risk management. Types of institutional-level support that can help include (1) subsidies for startup costs of leasing operations to help offset the higher transaction cost and risk of operating in rural areas; (2) funding to establish links between commercial providers and community-based or nonprofit organizations to increase scale; (3) technical support to leasing companies; and (4) provision of equity, loans, or guarantees to expand rural outreach.

A wide range of organizations—leasing companies, banks, financial cooperatives, microfinance organizations, and equipment-selling companies—could benefit from such support. Institutional-level support can include capital support when access to long-term funds is a critical constraint. Capital support combined with technical assistance can help leasing firms develop access to sustainable sources of capital. ■

For further reading: IFC (International Finance Corporation), *Leasing in Development: Guidelines for Emerging Economies* (Washington, DC, 2009), available at www.IFC.org; A. Nair and R. Kloeppinger-Todd, "Buffalo, Bakeries, and Tractors: Cases in Rural Leasing from Pakistan, Uganda, and Mexico" (Washington, DC: World Bank, 2006), available at www.worldbank.org/rural; A. Nair, R. Kloeppinger-Todd, and A. Mulder, "Leasing: An Underutilized Tool in Rural Finance," World Bank Agricultural and Rural Development Discussion Paper No. 7 (Washington, DC: World Bank, 2004), available at www.worldbank.org/rural; G. D. Westley, *Equipment Leasing and Lending: A Guide for Micro-Finance*, Best Practice Series (Washington, DC: Inter-American Development Bank, Sustainable Development Department, 2003).

***Ajai Nair** (anair@worldbank.org) is program coordinator of the Agriculture Finance Support Facility in the Agriculture and Rural Development Department of the World Bank. Azeb Fissha, consultant, provided research and editorial support.*

INTERNATIONAL FOOD POLICY RESEARCH INSTITUTE
Supported by the CGIAR
IFPRI® www.ifpri.org

THE WORLD BANK
www.worldbank.org

2020 VISION FOR FOOD, AGRICULTURE, AND THE ENVIRONMENT

Determinants of Microcredit Repayment in Federations of Indian Self-Help Groups

YANYAN LIU AND KLAUS DEININGER

FOCUS 18 • BRIEF 7 • JULY 2010

Since the establishment of the Grameen Bank in Bangladesh in 1976, microfinance has boomed. As of December 31, 2007, 3,552 microcredit institutions had reached 154 million clients worldwide, about 106.6 million of whom were among the poorest when they took their first loan. Such expansion can be at least partly attributed to the widely adopted practice of group lending in microfinance programs. In contrast to individual lending, group lending (or joint liability) grants a loan to a group of borrowers, and the whole group is liable for the debt of any individual member in the group. This practice allows microfinance programs to rely mainly on accountability and mutual trust among group members rather than financial collateral to insure against default. Given that the poor often lack appropriate financial collateral, group lending programs offer a feasible way of extending credit to poor people who are usually kept out of traditional banking systems.

There is considerable debate about whether such groups can be sustainable, achieving sound repayment performance while serving poor borrowers. The factors affecting repayment performance are thus of great policy relevance. This brief examines whether and how much repayment is affected by three factors: the source of the loan, groups' provision of public goods in the form of insurance substitutes, and the monitoring and repayment rules of the federations of groups. The data come from more than 2,000 self-help groups (SHGs), federated in 299 village organizations in the Indian state of Andhra Pradesh. The SHGs under study were supported by a large World Bank program called the Indira Kranti Patham (IKP) program, with a cost of US$260 million. The program has been replicated in other states in India and may be replicated in other countries. A better understanding of factors influencing repayment will therefore help improve the performance and advance of the program.

Background of the IKP program

Building on Andhra Pradesh's tradition of SHGs, the IKP program was launched in October 2000 to promote the formation of new groups and to strengthen existing ones. A typical program SHG consists of 10–20 women members who meet regularly to discuss social issues and engage in social activities. During these meetings each member deposits a small thrift payment into a joint bank account. Once enough savings have been accumulated, group members can apply for internal loans that draw on accumulated savings at an interest rate to be determined by the group. Once the group establishes a record of internal saving and repayment, it becomes eligible for loans through a commercial bank or IKP program funds.

An important component of the program is to support the federation of SHGs at the village and *mandal* (block/county) level through formation of village organizations and county organizations. The purpose of federation is to capitalize on economies of scale in capacity building, credit, and insurance and to ensure that public programs reach the poor. Although IKP program funds were initially made available to SHGs, they were shifted to village organizations and later to county organizations as soon as these were established.

The survey

Data for this study come from a survey of 299 village organizations conducted by the World Bank in 2006. This brief investigates 3,350 expired loans made to members of 2,147 SHGs. In the survey, all loans taken by each member SHG in the village organizations between June 2003 and June 2006 were recorded from account books of each organization. The study period started after the majority of village organizations were formed and coincided with a major drive for SHG formation. Of the 40 million rupees (about US$1 million) of aggregate loan principal, about 60 percent of the funds were provided by the IKP program, with the rest of the funds coming from banks, SHGs, and other sources. Only 63 percent of loans from the IKP program were fully repaid, compared with 87 percent repayment for bank loans and 89 percent repayment for internal loans.

The survey provides information on loan terms (size, source, length, interest rate, and repayment frequency), SHG characteristics (size, age, and membership composition), and village organizations' monitoring and repayment rules. These rules differ in four key dimensions:

- Delinquency management policies. These policies include fees to SHGs that miss an installment and loan recovery committees to monitor SHGs' creditworthiness (through a rating system, for example). Both would likely increase repayment probability.
- Monitoring of SHGs' financial affairs. Here, the study looks at three variables: whether the village organization (1) regularly inspects member SHGs' books at monthly meetings; (2) employs a trained bookkeeper; and (3) regularly audits members' books. Again, all of these steps should help reduce defaults.
- The extent to which the village organization provides public goods. The study considers whether in-kind rice credit and marketing services are provided. The in-kind rice credit is a program whereby the village organization acquires subsidized rice in bulk under the public distribution scheme and makes it available to SHG members as an in-kind credit, with any savings from the bulk purchase passed on to members in the form of lower prices. Marketing services are the collective activities that help SHG members gain access to markets—for example, buying and selling in bulk to obtain more favorable prices or to reduce transaction costs. Because such benefits can be cut off in case of default, they should enhance repayment incentives, especially when alternative sources for these benefits are unavailable.
- The extent to which SHGs are required to deposit regular thrift payments with the village organization. The village

organizations' collection of thrift from member SHGs provides cash collateral that can be withheld in case of default and thus should increase repayment incentives.

In the sample of 299 village organizations, 36 percent applied a sanction for SHGs that miss an installment, 41 percent had a loan-recovery committee, 35 percent provided in-kind consumption credit, 25 percent provided marketing services, 47 percent collected thrift from their member SHGs, 82 percent employed trained bookkeepers, 37 percent of the SHGs in the sample were regularly audited, and 23 percent presented their books at village organization meetings.

Factors influencing loan repayment

The model used to estimate the effects of various factors on repayment shows that monitoring and loan recovery arrangements are highly significant, both statistically and economically. Regular audits, checking of SHG books at village organization meetings, and depositing of SHG savings with the village organization are estimated to increase the probability of full repayment by 8.3, 9.5, and 20 percentage points, respectively. Although the village organization's involvement in marketing has no impact on repayment, in-kind consumption credit is predicted to increase the probability of full payment by 12.7 points, suggesting that non-economic benefits from credit groups increase repayment incentives. This finding also implies that village organizations are better positioned to help smooth consumption and address credit market imperfections than to intervene in output markets.

The results also suggest that SHGs are more likely to fully repay loans from banks—by 18.6 points according to the estimate—than loans from the IKP program. The program's lower repayment rate points to limits in village organizations' credibility, possibly because of their relatively recent establishment. High installment frequency has an almost equally large effect (15 points), consistent with the notion that frequent small installments enhance repayment performance for households with credit constraints. As have other studies, this study found that full repayment is less likely for loans with longer duration and, less significantly, higher interest.

Other studies have found mixed evidence on the impact of group characteristics, but the results of this study suggest that the probability of repayment increases with the size of the group up to about 14 members and decreases thereafter. In contrast, the probability of repayment decreases with the length of time the group has been in operation up to about five years. Although groups with a high percentage of poor individuals show lower rates of full repayment, the magnitude is small: a 10-point increase in very poor members would reduce full repayment by only 1.7 points. Here the trade-off between sustainability and service to the poorest is much smaller than suggested by some other studies. Neither caste composition nor homogeneity has a significant impact on repayment.

Summary and policy implications

In contrast to most existing literature that studies the effects of group and individual attributes on loan repayment in microcredit groups, this study investigates the effects of exogenous monitoring and loan recovery arrangements, together with loan and group characteristics. Because banks and others can provide microfinance institutions with additional resources contingent on adoption of certain minimum rules, the findings from this study could be of great practical relevance. The results highlight the following four policy implications:

1. Repayment rates are significantly lower on loans originating in externally provided grant resources managed by village organizations. This finding highlights the need for further inquiry on why this is the case and how to improve the repayment performance of loans from grant resources.
2. Among SHGs, external management policies (such as regular monitoring and audits and in-kind consumption credit) and loan terms (group savings deposits with the lender, frequency of repayment) appear far more important to full repayment than group characteristics such as the poverty level of members. This result suggests that, in this context, even groups composed of very poor borrowers can achieve high repayment rates if village organizations adopt proper rules and management practices. Furthermore, SHG federations and other external group supervisors should consider implementing the management policies that can encourage full repayment.
3. Third, the results suggest that the optimal size of a group is about 14 members. This finding can provide some guidance in group formation. ■

For further reading: C. Ahlin and R. M. Townsend, "Using Repayment Data to Test across Models of Joint Liability Lending," *Economic Journal* 117, no. 2 (2007): F11–51; R. Cull, A. Demirguc-Kunt, and J. Morduch, "Financial Performance and Outreach: A Global Analysis of Leading Microbanks," *Economic Journal* 117, no. 2 (2007): F107–33; M. Sharma and M. Zeller, "Repayment Performance in Group-Based Credit Programs in Bangladesh: An Empirical Analysis," *World Development* 25, no. 10 (1997): 1731–42.

Yanyan Liu *(y.liu@cgiar.org) is a research fellow in the Markets, Trade, and Institutions Division of the International Food Policy Research Institute (IFPRI).*
Klaus Deininger *(kdeininger@worldbank.org) is a lead economist at the World Bank.*

This brief is based on Klaus Deininger and Yanyan Liu, "Determinants of Repayment Performance in India Microcredit Groups," World Bank Policy Research Working Paper No. 4885 (Washington, DC: World Bank, 2009).

INTERNATIONAL FOOD POLICY RESEARCH INSTITUTE
Supported by the CGIAR
IFPRI® **www.ifpri.org**

THE WORLD BANK
www.worldbank.org

M-PESA: Finding New Ways to Serve the Unbanked in Kenya

SUSIE LONIE

2020 VISION FOR FOOD, AGRICULTURE, AND THE ENVIRONMENT

FOCUS 18 • BRIEF 8 • JULY 2010

Over the past three years, payment strategies for emerging markets have been revolutionized by the advent of a simple cell-phone-based payment service in Kenya called M-PESA ("M" for "mobile" and "pesa" for "money"). From a small-scale pilot program in 2006, M-PESA has become an outstanding success in Kenya; customer response has been unprecedented. Currently, more than 9 million Kenyans use M-PESA to perform tens of millions of transactions every month throughout the country. Although this success has led to new opportunities, it has also brought about many unforeseen challenges.

What is M-PESA?

Vodafone, the world's leading international mobile communications group, based in the United Kingdom, originally developed M-PESA with funding from the Department for International Development (DFID) as a pilot program to extend the growth of financial markets to the unbanked (people without access to conventional banking services) in East Africa. In March 2007, M-PESA was launched in Kenya in partnership with Safaricom, Kenya's leading mobile telecommunications company. It quickly became clear by the demand from the unbanked that this cell-phone-based, money-transfer business was a welcome commercial opportunity across Africa and elsewhere.

In emerging economies, it is common for some wage earners to work away from home and send domestic remittances back to their extended families in rural areas. The initial focus of M-PESA, therefore, was to enable these workers to send money home via faster, safer, and more affordable means than those previously available to them. M-PESA allows customers to send money home (and make a variety of other payments, described below) without a bank account. People can begin using the system simply by registering for free at certified M-PESA agents, which include retailers such as supermarkets, gas stations, and shops that sell prepaid airtime cards. In fact, several banks have even become M-PESA agents. Customers can use cash to "buy" electronic money (e-money) from an agent, then use their phones to perform financial transactions (for example, to send money to another person or buy additional airtime). The e-money can also be converted into cash by selling it back to an agent. Agents are paid a commission for providing cash-in and cash-out services and for registering customers.

Transaction values are typically low; M-PESA moves smaller amounts of money than banks would normally service. As M-PESA gains acceptance, however, it is also becoming attractive to people who already have bank accounts as a way to pay out wages to, for example, tradesmen and household staff—who are, of course, M-PESA's targeted customers.

M-PESA: Beyond "sending money home"

There is no doubt that giving M-PESA customers the ability to send money home was a feature that filled a gap in the market. Within two years, M-PESA had become the most frequently used money-transfer mechanism (see Figure 1 in Appendix A).

The need for fast, safe money transfers, particularly to those in rural communities, is obvious; and the benefits have been much reported in the Kenyan press and by nongovernmental organizations. However, by extending functionality and thinking laterally, M-PESA has been expanded to further serve the unbanked of Kenya.

Bill payments

The option for customers to pay their bills via cell phone was recently added to the M-PESA menu. Designed to allow people to pay their regular bills—such as utilities, school fees, and rent—this feature has become a means of payment collection for many other businesses as well. Of particular relevance to rural communities, even the provision of clean drinking water has been improved through the use of M-PESA. Grundfoss, a Danish company, developed an entrepreneurial solar-powered metering system to pump clean water from boreholes into rural areas. Pumps can be paid for by a "smart card" (a prepaid card with a memory chip), but finding means to apply credit to these smart cards in the rural areas where water pumps were situated was a problem—until M-PESA arrived. Customers sent money to Grundfoss using the "Pay Bill" feature, and their smart card was automatically credited and ready to use. This payment system is now available wherever Grundfoss water pumps are deployed.

Additionally, several microfinance institutions (MFIs) are now using M-PESA's bill-paying feature for loan repayment collection. This eliminates the time loan recipients used to spend travelling to urban areas to deposit cash into their MFI bank accounts; this time can now be better spent attending to their farms or small businesses. Similarly, insurance and microinsurance premiums can now be paid using M-PESA.

Business payments

The "Business Payments" feature allows a business to pay a number of customers or employees through their M-PESA accounts. This service was originally introduced at the request of Safaricom's temporary staff working in rural areas. These low-income workers previously had to travel to a Safaricom office in the nearest town to pick up their paychecks and deposit them into bank accounts; it was a time-consuming activity at best. Now they receive their wages directly through their M-PESA accounts. It has proven so popular that the organization recruited to provide M-PESA training to new agents around Kenya actually started to use the payment feature for its own staff expenses. Many other companies are now using M-PESA to pay field operatives working remotely from regional offices. Safaricom also recently offered shareholders the opportunity to receive their annual dividend payments via M-PESA; many thousands of Kenyans—who had become first-time shareholders when Safaricom issued public shares in 2008—accepted the offer.

Agent management tools

In rural areas, the only suitable outlets to become M-PESA agents are often small family-run stores. While it is not commercially feasible for M-PESA to have a direct business relationship with thousands of "Mom-and-Pop" shops across Kenya, a partnership of some kind would be advantageous to all parties. To resolve this conundrum, M-PESA created the aggregator model in which a distributor is appointed to recruit and manage agents in these small stores and in return gets a share of the M-PESA commission earned by that store.

To further accommodate the growing system, M-PESA had to address the issue that when an agent runs out of "float"—either e-money or cash—they cannot service their customers. With the huge expansion, getting more e-money into the system was a requirement and, oftentimes, a challenge since conventional banking takes some time to clear deposits before new e-money can be issued, during which time an agent who has run out cannot offer M-PESA. To reduce this problem, larger agents rich in cash or e-money were given the ability to act as agents to smaller shops. Thus, a smaller outlet can now buy or sell e-money from one of these larger, richer "super agents." Float management is a particular problem in rural areas where agents tend to net more withdrawals than deposits, or, in other words, they tend to be e-money rich but cash poor. The capability of these super agents has helped speed up the turnaround of e-money and cash, allowing small agents to have less money tied up in M-PESA and yet still have more float available.

Bank branches that were reluctant to become regular M-PESA agents for customers conducting small transactions have been happy to act as super agents for businesses operating with larger sums. This also gives these banks the opportunity to promote and sell their banking services to more businesses. Many small business owners acting as M-PESA agents now have their first ever bank account.

Banking services

M-PESA is now giving cell-phone users access to formal banking services. In May 2010 Safaricom and Equity Bank, a leading bank in Kenya, launched an initiative to offer every M-PESA user the opportunity to open a savings account. Customers use M-PESA to both deposit money into and withdraw money from their savings accounts. Called M-KESHO (*kesho* is Swahili for "tomorrow"), this service effectively gives millions of rural Kenyans access to banking services for the first time.

M-PESA: Perks and pitfalls

It became clear soon after M-PESA's launch that the service provided an effective and convenient means of making any sort of person-to-person money transfer, and subscriber numbers grew well beyond projections. Initial predictions estimated 320,000 users in the first year of trading: Nine months after launch, M-PESA registered its one millionth customer. It was an exciting success. As with any new business that unexpectedly finds itself growing much faster than anticipated, however, there were numerous implications, each with significant costs.

- Budget flexibility: The rapid growth required a significantly reworked budget. For example, customer acquisition costs money because agents have to be paid to register customers, and the cost of new SIM cards—which are free to new customers—needs to be covered. It takes time for new customers to become mature users and start generating revenue, so signing up more than 10,000 new customers per day had a serious impact on cash flow in the early life of the product.
- Customer support: M-PESA needed a significantly larger call center and a lot more customer service representatives than were originally anticipated.
- System capacity: At the time of M-PESA's launch, the system had a technical design that could cope comfortably with the original business case plus a sensible safety margin; this capacity was rapidly exceeded and had to be regularly expanded to include new features at significant expense.
- Managing agent demand: A sufficient number of agents had to be sought out, enrolled, and trained. As retailers and outlets came to understand the business opportunity, however, the situation reversed; their demand was such that the Safaricom sales office had to cater to the crowds of would-be agents. Extra staff members were also required to process applications and provide ongoing agent training.

Conclusion

As problems go, those associated with rapid growth are the best kind to have, but they are challenging nonetheless. Substantial costs were incurred far earlier than anticipated, pushing back the expected break-even date. Working on the basis that budget shortfall would soon be forgotten while unexpected customer growth would be remembered for years to come, M-PESA managed to secure additional funds. Now, alongside increasing demand, a critical mass of mature customers is growing to support the need for revenue. There is still much to do, but as M-PESA approaches its third birthday in Kenya, it is well prepared to tackle whatever comes next. ■

See more information at www.safaricom.co.ke/index.php?id=745.

Susie Lonie *(Susie.Lonie@vcontractor.co.za) is a mobile commerce expert and currently the executive head of financial services at Vodacom (Pty) managing the rollout of M-PESA in South Africa. Lonie managed the design, piloting, and implementation of M-PESA in Kenya.*

INTERNATIONAL FOOD POLICY RESEARCH INSTITUTE
Supported by the CGIAR
IFPRI® www.ifpri.org

THE WORLD BANK
www.worldbank.org

2020 VISION™ FOR FOOD, AGRICULTURE, AND THE ENVIRONMENT

Biometric Technology in Rural Credit Markets: The Case of Malawi

XAVIER GINÉ

FOCUS 18 • BRIEF 9 • JULY 2010

Identity theft is a common crime the world over. In developing countries, the damage caused by identity theft and identity fraud goes far beyond the individual victim, however, and ultimately creates a direct impediment to progress, particularly in credit markets. Recent research reveals that biometric technology can help reduce these problems.

A biometric is a measurement of physical or behavioral characteristics used to verify or analyze identity. Common biometrics include a person's fingerprints; face, iris, or retina patterns; speech; or handwritten signature. These are effective personal identifiers because they are unique and intrinsic to each person, so, unlike conventional identification methods (such as passport numbers or government-issued identification cards), they cannot be forgotten, lost, or stolen.

Recent advances in recognition technology coupled with increases in both digital storage capacity and computer processing speeds have made biometric technology (for example, ocular or fingerprint scanners) feasible in many applications, from controlling restricted building access to allowing more effective delivery of targeted government programs with large-scale identification systems, such as those being implemented in India by the Unique Identification Authority of India.

Biometric technology can also improve access to credit and insurance markets, especially in countries that do not have a unique identification system, where identity fraud—the use of someone else's identity or a fictitious one—to gain access to services otherwise unavailable to an individual is rather common. For example, lenders in Malawi describe past borrowers who purposefully defaulted then tried to obtain a fresh loan from the same or another institution under a false identity. And, although less common in developing countries because markets are less developed, the potential for sick individuals without healthcare coverage to use the insurance policy of a friend or relative does exist. The response of lenders and insurance companies has been to restrict the supply of such services to the detriment of the greater population, not just those people committing identity fraud.

In the case of credit, biometric technology can make the idea of future credit denial more than an empty threat by making it easier for financial institutions to withhold new loans from past defaulters and reward responsible past borrowers with increased credit. As a result of this inability to "cheat the system," individuals may take out smaller loans that they are able to repay or avoid borrowing altogether if they cannot pay back any debt. Borrowers may have greater incentives to ensure that production is successful, either by exerting more effort or choosing less risky projects, and—whenever production could cover the loan repayment—borrowers may be less likely to default intentionally or opportunistically.

To look at the impact of biometric technology, Giné, Goldberg, and Yang (2009) implemented a field experiment using 3,200 smallholder paprika farmers in four locations in Malawi who applied for an agricultural input loan in 2007. Farmers in the study were randomly allocated to either a control group or a treatment group; each member in the latter group had a fingerprint collected as part of their loan application and an explanation that this would be used to determine their identity on any future applications. (Fingerprint recognition was used instead of face, iris, or retina recognition because the technology has been commercially available since the early 1970s, and there is a highly competitive market for it. Therefore, it is inexpensive, well known, and widely used.) Both treatment and control groups were given a training session on the importance of credit history in ensuring future access to credit.

The study shows that within the subgroup of farmers who had the highest ex ante default risk, fingerprinting led to increases in repayment rates of about 40 percent. By contrast, fingerprinting had no impact on repayment for farmers with low ex ante default risk. These higher repayment rates are due to fingerprinted borrowers requesting smaller loan amounts to ensure they would be able to repay them and devoting more land and inputs to paprika, thus diverting fewer resources to other crops; the same cannot be said for their nonfingerprinted counterparts.

A rough cost–benefit analysis of the pilot experiment suggests that the benefits from improved repayment greatly outweigh the costs of biometric equipment and fingerprint collection, which accounts for basic training and the time it takes credit officers to collect biometric data. These costs, however, do not include a full implementation plan, which would likely require software integration, expanded data-storage facilities, upgraded equipment, and more in-depth staff training.

Challenges in the implementation of biometric systems

Despite the encouraging results from the pilot in Malawi and the success of biometric technology in controlled laboratory environments, there are still concerns and challenges when collecting and using such information in real life and when trying to establish an identification system at a national level.

- Not everyone can participate in a fingerprint-based identification system. Fingerprints can be unrecognizable due to cuts or burns. In addition, older individuals may have fingerprints that have worn with age, and the operation of fingerprint readers may be jeopardized due to arthritis. In some areas, especially those with past or present conflict, individuals may lack fingers altogether. In the most comprehensive study to test the process and customer attitude during the recording of biometric information, the United Kingdom passport service trial reports an enrollment success rate of 100 percent for the 9,250 nondisabled participants and 96 percent for the 750 disabled participants. In Malawi, only about 2 percent of the sample of 1,600 fingerprinted farmers had to have their left thumbprint recorded when the scanner failed to capture the required right thumbprint. This is surprising, as it turns out, because many Malawian farmers grow tobacco, which requires the heavy use of fingertips in the transplant of seedlings. Over the years, their fingerprint ridges may become too worn to be read or captured by a fingerprint scanner.

- The accuracy of biometric technology remains, to a large extent, untested. Biometric companies report very high accuracy rates from highly controlled trials that typically use artificially generated data. However, because the performance of a technology depends greatly on the context in which it is used, trials using real-life data are far less impressive. For example, the United Kingdom passport service trial reports that only 80 percent of the participants could be correctly verified by their fingerprints, and younger individuals were more successfully identified than older ones. In Malawi, however, everyone selected during demonstration sessions was correctly identified.
- Individuals may have a negative attitude toward providing their biometrics. People may be reluctant to place their fingers on scanners due to hygiene concerns. More importantly, there is the widespread public perception that fingerprinting is linked to the criminal justice process. Therefore, in conflict-affected countries that are stricken by ethnic infighting, individuals may refuse to provide biometrics for fear of persecution by authorities or others who could gain illegal access to such biometric records. The parliamentary debates concerning the United Kingdom's identification cards bill revealed that 55 percent of poll respondents thought the collection of biometric information was an infringement of civil liberties. The authors did not encounter any such resistance from farmers in Malawi, perhaps because the technology was very novel.
- The cost of collecting biometrics can be high. The estimates are sparse, and detailed cost-benefit analyses have not been systematically conducted. However, the costs of using different types of biometric technology—from basic fingerprinting techniques to voice- and iris-recognition software—can be prohibitively expensive. In India there are legitimate concerns that the costs of rolling out biometric technology may mean a huge opportunity cost for more than 700 million Indians living in poverty to receive social benefits. In the United Kingdom, a critical report by several researchers at the London School of Economics and Political Science found that the government underestimated the implementation of the Identity Cards Bill. The report suggests that the ten-year rollout would cost between 10.6 billion and 19.2 billion pounds (compared to the government estimate of 5.84 billion pounds over the same period), excluding public- or private-sector integration costs.
- Biometric technology is not infallible. While biometric identification systems can help combat identity theft, fraud, and money laundering, they are essentially technological applications and, as with any other technology, can be hacked or infiltrated. These systems therefore run the risk of having data fall into the wrong hands. Since biometric technology is only being piloted on a large scale in some pockets of the world at present, legitimate concerns on privacy do arise. For example, it is possible to imagine that identification-database workers will be threatened, blackmailed, and possibly corrupted. After all, the perpetrators of 80 percent of all computer security lapses are not hackers, but employees.
- It is important that a common platform be used if biometric data are merged with other datasets. Biometric data are stored in formats that may not be compatible with the information systems of other government agencies, so an effort must be made to have compatibility if biometrics are to serve as the basis for a national identification system.

Conclusion

Despite these concerns, biometric technology presents an exciting and innovative opportunity for increased access to financial markets and better delivery of social assistance programs such as conditional cash transfers, aid distribution, or subsidized inputs or commodities. Whether it can be scaled up effectively and used to resolve identification and authentication issues is a challenge that requires more research. ■

For further reading: X. Giné, J. Goldberg, and D. Yang, "Identification Strategy: A Field Experiment on Dynamic Incentives in Rural Credit Markets," mimeo (Washington, D.C.: World Bank, 2009); U.S. General Accounting Office, *Using Biometrics for Border Security* (Washington, D.C.: 2002); London School of Economics and Political Science (LSE) Identity Project, www.identityproject.lse.ac.uk.

Xavier Giné *(xgine@worldbank.org) is a senior economist in the Research Development Group at the World Bank.*

INTERNATIONAL FOOD POLICY RESEARCH INSTITUTE
Supported by the CGIAR
IFPRI® **www.ifpri.org**

THE WORLD BANK
www.worldbank.org

2020 VISION FOR FOOD, AGRICULTURE, AND THE ENVIRONMENT

Credit Risk Management in Financing Agriculture

MARK D. WENNER

FOCUS 18 • BRIEF 10 • JULY 2010

Agriculture is an inherently risky economic activity. A large array of uncontrollable elements can affect output production and prices, resulting in highly variable economic returns to farm households. In developing countries, farmers also lack access to both modern instruments of risk management—such as agricultural insurance, futures contracts, or guarantee funds—and ex post emergency government assistance. Such farmers rely on different "traditional" coping strategies and risk-mitigation techniques, but most of these are inefficient. Formal and semiformal arrangements—such as contract farming, joint-liability lending, and value-chain integration—have arisen in recent decades, but they too are limited and can be very context sensitive. One consequence of inadequate overall financial risk management is that farmers in general face constrained access to formal finance. The smaller the net worth of the farm household, the worse the degree of exclusion.

Formal lenders avoid financing agriculture for a host of reasons: high cost of service delivery, information asymmetries, lack of branch networks, perceptions of low profitability in agriculture, lack of collateral, high levels of rural poverty, or low levels of farmer education and financial literacy. But, predominantly, bank managers around the world say they will not finance agriculture because of the high degree of uncontrolled production and price risk that confronts the sector. A farmer can be an able and diligent manager with an excellent reputation for repayment, guaranteed access to a market, and high-quality technical assistance, but an unexpected drought or flood can force him or her to involuntarily default. In emerging countries with fair to high levels of agricultural market and trade integration, large commercial farmers may escape this predicament because they have the ability to purchase insurance, engage in price hedging, obtain financing overseas, or liquidate assets quickly in the event of a default. Consequently, formal lenders tend to overemphasize the use of immoveable collateral as the primary buffer against default risk, which means they provide services to a limited segment of the farm population. Small- and medium-sized farmers, who constitute the vast majority of farm operators, often do not have secured-title land, which is the preferred type of collateral; if they do, its value may be insufficient to cover the loan in question. Even if farmers have sufficient titled land to collateralize loans, they may refuse low-interest formal loans and assume high-interest informal ones that have no collateral requirements instead. They may also use savings to finance agricultural production because they are averse to risking their most prized possession—land. The result is limited supply or access to formal agricultural financing, even though much of the population of Sub-Saharan Africa and South Asia is rural and depends on agriculture and livestock rearing for their main livelihood activities.

Typical risk-management mechanisms in rural financial intermediaries

In developing countries, formal and semiformal rural financial intermediaries have limited or nonexistent means to transfer credit risk to third parties through, for example, portfolio securitization or credit insurance, which were common in mortgage and consumer finance markets in developing countries prior to the 2008 financial crash. If more farm borrowers held agricultural insurance policies, this could serve to reduce credit risk for financial institutions, but agricultural insurance markets are grossly underdeveloped in middle- and low-income countries. For example, agricultural premiums totaled US$18.5 billion worldwide in 2008, but the United States and Canada accounted for 62 percent of the premium volume. Latin American, Asian, and African regions, home to most of the lower-income countries, accounted for 21 percent, or US$3.88 billion. Moreover, the leading countries in terms of agricultural insurance development—the United States, Canada, and Spain—all depend on heavily subsidized schemes that would be difficult to replicate in other places.

Thus, most of the strategies available to financial intermediaries in developing countries involve coping with and absorbing credit default risk. There are two broad means of evaluating creditworthiness: appraisal of repayment capacity and asset-backed lending. The former approach focuses on analyzing the debt-paying capacity of a potential borrower using either human experts or statistical models, while the latter focuses on the quality and quantity of assets that can be pledged as collateral and how quickly that collateral can be liquidated in the event of a default. Since titled assets are scarce outside of large farms and extensive databases on farm enterprises rarely exist in developing countries, the following represent the four credit risk-management techniques used successfully by rural financial intermediaries.

Expert-based credit evaluation systems: Trained credit officials conduct financial analysis of the client, focusing on household cash flow, market situation, assessment of managerial or entrepreneurial ability, and reputation. Institutions can have centralized or decentralized systems to approve client requests as long as both systems include performance incentives for and investments in staff members, who should be recruited from the region of operations. To quickly determine client willingness to repay loans, staff members need access to credit bureaus or borrowers' utility bill payments. Agriculture requires a wide range of experts since it is such a heterogeneous field; therefore, an expert-based evaluation system is expensive to both develop and maintain.

Portfolio diversification: In order to dilute risk, intermediaries consciously seek to diversify the agricultural loans approved by geographic region, commodity, and type of household. This technique can be implemented only by large institutions that operate in more than one agroclimatic zone, however.

Portfolio exposure limit: Because agricultural lending is risky and expensive, high-performing financial intermediaries tend to limit exposure to agriculture in their loan portfolio. For example, recent survey data in Latin America found that the average share is less than 40 percent. The smaller the share agriculture has in a total loan portfolio, the less vulnerable the institution is to systemic external shocks that could severely depress earnings performance and the more cross-subsidization can occur. High-margin financial products—such as consumer finance and urban microfinance—can compensate for lower profit margin products, such as agricultural loans.

Excessive provisioning: The last line of defense is called "loan loss provisioning," meaning an internal absorption of credit risk. Adequate provisioning according to a risk-classification scheme helps to protect the intermediary from liquidity and capital adequacy crises. Some leading agricultural lenders in Latin America, for example, provision from 121 to 260 percent of doubtful loans. Heavy provisioning, however, clearly constrains the volume of lending, ability to make a profit, and client outreach potential.

Implications for managers of financial institutions and public policymakers

There are numerous implications of these credit risk-management techniques. First, the credit risk evaluation systems are labor intensive with high costs, which, in turn, contribute to high lending interest rates. Public-sector policymakers need to understand this, so they avoid imposing interest rate ceilings or forcing publicly owned banks to charge interest rates that are lower than their true operating costs because results would then be counterproductive. Additionally, fewer intermediaries would be willing or able to serve the sector. Therefore, both policymakers and managers should focus on developing and implementing institutional innovations—such as credit bureaus, applications of information and communication technology, and delegated agent models of service delivery—that will reduce overall operating costs.

Second, agricultural lending cannot be the primary type of lending unless robust risk-transfer techniques (for example, insurance, futures, and securitization) become more commonplace. In place of land, alternative forms of collateral—including warehouse receipts, accounts receivable, equipment, and standing crops or livestock—should be more widely accepted. Improved contract enforcement should be aggressively promoted as well. These developments would all serve to lower lender risk. Many of these innovations and institutional developments require legal and regulatory reforms, modernization of property registries, investments in information infrastructure, and massive education efforts.

Third, the majority of institutions involved in agricultural lending are small and unregulated. They are using adapted microcredit-lending technologies that do not fully meet the needs of farmers, especially those needs regarding loan term and repayment frequencies. These shortcomings pose default risks in and of themselves. The larger institutions that have the scale and scope tend not to enter into agricultural lending because they do not have the strategic commitment, proper staff, or branch networks. Donors and governments can play a vital role in assisting these smaller institutions to grow, consolidate, and eventually merge. They can also help rural financial intermediaries with liability diversification through mobilization of savings, access to capital markets, and the provision of long-term lines of credit that could facilitate more term lending. Nevertheless, donors and governments must price the discount line of credits in a manner that will not undermine savings mobilization.

Conclusion

In short, risk management needs to improve dramatically so that agricultural finance can flourish. Strides have been made in recent years in reducing information problems and transaction costs through, respectively, peer-group lending and a greater reliance on information and communication technology. Uncontrollable risk, however, continues to be a major impediment to the development of more efficient rural financial markets. Renewed private-public sector efforts and higher amounts of investments will be required at various levels to address these issues. At the farmer level, governments need to spur the rebuilding of farm extension services, while farmers need to become more financially literate and save more so they can retain some of the risks. Governments, donors, and insurance companies need to collaborate in the development of yield-insurance products that are inexpensive, sustainable, and appropriately designed. Governments, commodity exchanges, and financial institutions likewise need to collaborate in developing futures, structured finance products, and other hedging instruments to reduce price risk.

At present, the lack of high-quality weather data, inadequate distribution of weather stations, limited supply of people with risk-modeling capabilities and expertise in agricultural risk management, small capital markets, and weaknesses in regulatory and legal infrastructure hamper the pace of progress. Since the depth and efficiency of financial markets are highly correlated with the speed of overall economic development, innovative methods of improving rural financial services will be critical in facilitating and sustaining any marked improvement in rural welfare. ■

For further reading: H. Bhattacharya, *Banking Strategy, Credit Appraisal and Lending Decisions: A Risk-Return Framework*, (New Delhi, India: Oxford University Press, 1996); J. B. Caoutte, E. I. Altman, and P. Narayanan. *Managing Credit Risk: The Next Great Financial Challenge*, (New York: John Wiley and Sons, Inc., 1998); C. Trivelli, and A. Tarazona, *Riesgo y Portafolios Agropecuarios: Lecciones desde la Experiencia de Instituciones Financieras de América Latina*, Documento de Trabajo 151 (Lima, Perú: Instituto de Estudios Peruanos, 2007), www.iep.org.pe/textos/DDT/DDT151.pdf; M. Wenner, S. Navajas, C. Trivelli, and A. Tarazona, *Managing Credit Risk in Rural Financial Institutions in Latin America*, Sustainable Development Department Best Practices Series MSM 139 (Washington, D.C.: Inter-American Development Bank, 2007); World Bank, *Doing Business* database, www.doingbusiness.org/economyrankings.

***Mark D. Wenner** (markw@iadb.org) is a lead financial specialist in the Capital Markets and Financial Institutions Division of the Inter-American Development Bank.*

INTERNATIONAL FOOD POLICY RESEARCH INSTITUTE
Supported by the CGIAR
IFPRI® **www.ifpri.org**

THE WORLD BANK
www.worldbank.org

New Approaches for Index Insurance: ENSO Insurance in Peru

Jerry R. Skees and Benjamin Collier

2020 Vision for Food, Agriculture, and the Environment

Focus 18 • Brief 11 • July 2010

The El Niño Southern Oscillation (ENSO) is a climate event associated with warming sea surface temperatures in the Pacific Ocean. In years of extreme El Niño events, areas in northern Peru experience catastrophic flooding. As of 2010, it is possible for stakeholders in northern Peru to purchase a new form of insurance that pays out just as flooding begins and stakeholders begin incurring extra costs and consequential losses. Given the high basis risk associated with selling index insurance to households, this insurance is designed for firms and institutions that serve households that are highly exposed to El Niño. ENSO insurance is sold by a Peruvian insurance company, and a major global reinsurer carries most of the risk. This new insurance product is the first insurance to use sea surface temperature as the proxy for catastrophic losses and also the first regulated "forecast index insurance" product in the world. This innovation could enhance progress in developing index-based insurance products for extreme weather events.

Recent years have seen a growing number of pilot tests of index insurance for weather risk, motivated by an increased understanding of how natural disasters affect developing countries. Beyond immediate suffering (including deaths, destroyed assets, and lost income), disasters have troublesome indirect effects: economic growth can be disrupted, the poor are thrust into permanent poverty traps, and the mere presence of these risks constrains access to financial services and causes many decisionmakers to pursue low-return, low-risk strategies that impede economic progress.

Much of the development of index insurance focuses on agriculture, because activities associated with agriculture remain the primary livelihood strategies for the rural poor in developing countries. Thus far, most index insurance pilots have involved products targeted at households—that is, micro-level products. Index insurance uses an objective measure (an index) of a natural event known to cause losses (such as excess rain, high river levels, or extreme sea surface temperatures). Using an index as the proxy for loss dispenses with expensive loss assessments. Furthermore, use of an index diminishes moral hazard and adverse selection, problems that plague traditional forms of insurance. Given these advantages, index insurance may be well suited to developing countries where data are sparse and delivery of financial services to smallholder households increases the per-unit cost of traditional insurance.

Despite the promise of index insurance, uptake by smallholder households is slow. Presently, index insurance may be better suited for risk aggregators—that is, groups or institutions that aggregate the risk of households either through the services they provide or through informal risk-sharing arrangements (for example, agricultural lenders, firms in the value chain, and farmer associations). Focusing first on risk aggregators should also help build linkages and sustainable products that will directly serve smallholder households.

Index insurance and correlated losses

As a precondition of index insurance, losses created by the natural disaster to be insured must be strongly correlated—that is, a large number of individuals must suffer losses at the same time. Given that many individuals suffer a loss at the same time, risk aggregators will also suffer serious losses. Thus, correlated losses from natural disasters constrain the development of credit markets for the rural poor, particularly for those involved in agriculture. Lenders cannot absorb the risk exposure of a large number of borrowers who may be unable to pay off loans after a major natural disaster.

Likewise, an insurer deciding to write any form of insurance against extreme weather events must have a means to transfer these risks—generally through a global reinsurer. Insurers in developing countries often find obtaining access to reinsurance markets difficult. If the index being used is fully transparent, the global reinsurer is more likely to feel comfortable with the systems used to estimate the index. This is certainly the case for ENSO measurements, which have been developed over more than 50 years by the U.S. National Oceanic and Atmospheric Administration (NOAA).

Extreme weather events such as drought and flooding can also have associated consequential losses that extend beyond traditional crop insurance, which pays for losses of a specific crop. For example, in a number of African countries, where owning livestock is a form of savings, extreme droughts compel large numbers of farmers to sell their livestock at the same time. Distressed sales of livestock on local markets depress local prices, compounding losses. Floods and droughts also generally influence the quality of crops, not just yields. Moreover, risk-management strategies to diversify cropping enterprises can quickly prove ineffective if droughts or floods negatively affect all of the crops at the same time.

ENSO insurance as a form of business-interruption insurance

In Peru, where ENSO insurance is being tested, the consequential losses and problems associated with extreme rainfall and catastrophic flooding are enormous—crops are lost, trees die, soils wash away, transportation systems break down, incidence of disease (such as malaria) increases, and markets are destroyed. When individuals and local markets suffer in this fashion, firms in the value chain and the financial sector also suffer.

In the extreme El Niño years 1983 and 1998, the volume of water in the Piura River was about 40 times greater than normal. Although Piura was among the worst-affected areas, a number of other regions in northern Peru were also severely affected. In 1998, with a clear indication that El Niño was coming, farmers simply did not plant crops, resulting in a 27 percent drop in fertilizer sales in northern Peru. Agricultural lending was growing at a significant pace before the 1997–98 El Niño, but that growth came to a halt after the event. Microfinance institutions (MFIs) had a significant increase in problem loans. Because of the 1998 El Niño, the default rate on agricultural loans increased from about 8 percent of all agricultural loans to more than 18 percent for MFIs operating in the region of Piura. Loan default is defined as loans that were 60 days late or more in payments. Once loans fall into this category, the probability of collection is nearly zero. Additionally, member deposits and savings—the major sources of capital for the MFIs— declined by roughly 15 percent as people withdrew funds to cope with the problems created by the event. It took at least three years to recover from the compounded problems of loan defaults, loan restructuring, and savings and member deposit withdrawals.

ENSO insurance was presented to the Peruvian insurance regulators as a form of business-interruption insurance designed to pay for consequential losses and extra costs linked to extreme flooding, which is highly correlated with ENSO. ENSO insurance fits well in a class of insurance called "contingency insurance." Contingency insurance is intended to protect policyholders against a variety of consequences associated with a specific event; these consequences can include loss of assets, losses in normal business revenues, and increased costs associated with addressing the event. Experience in Peru suggests that formulating index insurance as contingency insurance against a natural disaster has potential applications in many regions of the world highly exposed to severe weather risks such as drought or flood.

The ENSO insurance uses the monthly sea surface temperature for ENSO Region 1.2 (0–10° South, 80–90° West), measured and reported by the NOAA Climate Prediction Center. The basis for payment is the average of two months—November and December. Three contracts are available with three different thresholds where payments begin (23.4, 24.0, and 24.5 degrees Celsius); each of these contracts reaches a maximum when the measure reaches 27 degrees. The payout function is linear. Indemnity payments are made in early January, just as flooding begins, and flooding continues from February to April.

Indemnity payments are made by multiplying the payout rate times the sum insured, which is selected by the insured party. A risk assessment that estimates the largest losses that may occur under the worst flooding event is likely the best starting point for selecting a sum to be insured. Prudent buyers of insurance will be more likely to select a value that is less than these estimates, given the expense of this type of insurance and the fact that they have other risk-management mechanisms that can be blended with the ENSO insurance in an optimal fashion.

Since the ENSO insurance pays before the catastrophe, educational efforts have focused on helping people in the target markets understand how to use the extra cash to mitigate the impending crisis. Farmers' associations in remote regions of Piura have expressed an interest in using the funds to clear drainage systems because heavy rains associated with ENSO increase the likelihood that drainage systems will clog. Lenders are interested in using payments to ease the liquidity crisis and associated cost. Those in the value chain are interested in smoothing their losses and maintaining their specialized workforce when revenues are temporarily low because of El Niño. Finally, ENSO insurance is also being offered to local and regional governments to provide ready cash to mitigate some problems associated with catastrophic flooding.

To begin, the insurance company is offering ENSO insurance only to highly exposed risk aggregators. Demand from other firms and institutions that are exposed to El Niño risk will then drive the expansion of this market. Anecdotal evidence points to substantial interest in ENSO insurance. After some initial press releases on the product, the insurer was inundated with calls from a variety of firms and institutions interested in the product. At this stage, ENSO insurance is not being made available to smallholder households. The product can, however, be tied to other financial services in a fashion that gives smallholders greater access to these services at better prices.

Conclusion

El Niño events affect many regions of the world. The most dramatic effects probably occur in Peru and Ecuador, but El Niño affects other countries in South, Central, and North America as well as in Southeast Asia and East Africa. In some regions, El Niño is associated with flooding, and in others it is associated with drought. Although no other region may have as strong a correlation between sea surface temperature and flooding as northern Peru and southern Ecuador, this project may increase awareness and lead to new thinking and opportunities regarding the potential for forecast index insurance and the relationship between natural disaster risk and oceanic oscillations such as ENSO. ■

For further reading: J. R. Skees, J. Hartell, and A. Murphy, "Using Index-based Risk Transfer Products to Facilitate Micro Lending in Peru and Vietnam," *American Journal of Agricultural Economics* 89 (2007): 1255–61; J. R. Skees, "Challenges for Use of Index-based Weather Insurance in Lower-Income Countries," *Agricultural Finance Review* 68 (Spring 2008): 197–217; J. R. Skees and B. J. Barnett, "Enhancing Micro Finance Using Index-based Risk Transfer Products," *Agricultural Finance Review* 66 (2006): 235–50.

Jerry R. Skees *(jskees@uky.edu) is president of GlobalAgRisk and the H. B. Price professor of policy and risk in the Department of Agricultural Economics at the University of Kentucky.* ***Benjamin Collier*** *(benjamin@globalagrisk.com) is an employee of GlobalAgRisk and a PhD candidate in the Department of Agricultural Economics at the University of Kentucky.*

INTERNATIONAL FOOD POLICY RESEARCH INSTITUTE
Supported by the CGIAR
IFPRI® **www.ifpri.org**

THE WORLD BANK
www.worldbank.org

2020 VISION FOR FOOD, AGRICULTURE, AND THE ENVIRONMENT

Microinsurance Innovations in Rural Finance

MARTINA WIEDMAIER-PFISTER AND BRIGITTE KLEIN

FOCUS 18 • BRIEF 12 • JULY 2010

Poor people in developing countries are vulnerable to a broad range of shocks that affect their livelihoods, including illness, accidents, and death as well as loss of assets such as animals, crops, and machinery. The poor are still predominantly rural, and their vulnerability is even higher than that of their urban peers. Health facilities are less available and less well equipped in rural areas; water, sanitation, roads, and telecommunication are less developed; and people are less educated and not as aware of risk-mitigation mechanisms. Given the rural character of poverty in many countries, poverty reduction remains strongly connected to agricultural development, and sustainable agricultural development depends on well-organized risk mitigation. One important tool for mitigating risk is microinsurance.

The International Association of Insurance Supervisors (IAIS) defines microinsurance as "insurance that is accessed by the low-income population, provided by a variety of different providers but run in accordance with generally accepted insurance practices (including the IAIS Insurance Core Principles)." It differs from traditional insurance in that it is adapted to the circumstances of the poor: premiums are low, products have simple designs, it is offered through well-trusted and innovative channels, premium payments are flexible, and claims are settled promptly.

Microinsurance has the potential to enable the rural poor to mitigate the effects of shocks that threaten their lives, productivity, and assets. It can help prevent emergencies from depleting poor people's savings and other assets. Furthermore, it allows households to invest in high-risk, high-return activities by securing the lending risk for agricultural and other investments.

Financial sector reforms in many countries have begun to include insurance as an important pro-poor financial service along with other microfinance services such as savings, lending, and cashless payments. According to a study by the International Labour Organization, microinsurance in Africa almost doubled from 2006 to 2009. The survey shows that half of the schemes were growing faster than 30 percent a year between 2007 and 2008. Data on growth in rural areas, however, are not available.

Microinsurance in rural areas

Providing microinsurance in rural areas can be more difficult than in urban settings and requires some adaptations in terms of insurance products, risk carriers, delivery, and servicing. The characteristics of its market make sales and pricing more difficult. Because educational levels are usually lower in rural areas, low-income households' financial capability is weaker. Potential customers often distrust insurance. Affordability is another challenge because rural people rely on seasonal and generally low incomes. Demand is not known or understood, products are poorly designed, and if microinsurance is available at all, the selection of policies is limited. Systems are not adapted to manage many small transactions for premium collection, back-office administration, and claims management. Population density is often low and distances are far, making it more difficult to reach scale. Distribution can suffer from the lack of channels like banking outlets. Risk assessment is difficult because of a lack of mortality and morbidity data. All these factors make underwriting more expensive, and therefore, the mostly urban-based insurers are not ready to serve rural low-income markets. Rural providers are often small and informal, with the inherent challenges of such organizational forms, such as weak governance or limited range of products.

Experiences in providing microinsurance to the rural poor

Some microinsurance schemes tailored to rural areas provide useful lessons. For example, a pilot project in India, set up by the microfinance institution BASIX and a commercial insurer with the help of the World Bank, has been providing weather insurance for small farmers to improve their access to credit. This microinsurance scheme is based on a rainfall index. Payments are based not on individual loss adjustments—a costly undertaking not feasible in microinsurance—but rather on whether rainfall measured at a local weather station reaches a certain threshold. The insurance contracts are linked to credit because the insurance secures repayment of the loans.

At the outset, the project had to solve several problems like poor-quality weather data. Furthermore, the insurance provider strongly focused on raising awareness, capacity building, continuous product improvement, and quick payouts. Certain elements remain unresolved, such as basis risk (which occurs when the actual loss does not match the benefit because a payout is triggered but there is no loss or vice versa), high premiums, clients' difficulties understanding this complex product, and low demand due to these factors. Nevertheless, the Indian weather insurance market is growing strongly, and new microinsurance providers are entering the market.

Innovations are also taking place in the policy area. Insurance regulators across the globe are working to create an enabling framework for insurance products, delivery channels, and new providers. Brazil and Ghana are among the countries changing their regulations, and South Africa aims to integrate thousands of existing informal burial societies into the mainstream insurance sector.

The Philippines has already made such a regulatory move. In 2006 the Insurance Commission of the Philippines issued Insurance Memorandum Circular No. 9-2006, which provided a definition of microinsurance and spelled out the requirements for registering a microinsurance mutual benefit association. As a result, by the end of 2009, microinsurance mutuals covered 2 million policyholders. Now the Philippine government is pursuing a wider approach that aims to increase access to microinsurance products and services by including different kinds of actors. The country's National Microinsurance Strategy and regulatory framework focus on increasing private-sector participation in the provision of microinsurance and the mainstreaming of informal insurance.

Another groundbreaking example of dedicated microinsurance regulation can be found in India. India was the first country to pass regulations covering microinsurance products and agents in 2005. In addition, India made it compulsory for all insurers to provide microinsurance to the rural and social sectors (the "social sector" includes the unorganized sector, informal sector, economically vulnerable or backward classes, and other categories of persons in both rural and urban areas).

Although the evidence on requiring insurers to provide microinsurance is mixed (as insurers may consider the penalties a cost of doing business), this approach has stimulated a large number of pilot projects and provided useful lessons for the industry and for policymakers.

Today, insurance supervisors all over the world are expressing strong interest in sharing experiences and understanding the dynamics of an enabling regulatory and policy environment. In response to this interest, the IAIS, in partnership with the Consultative Group to Assist the Poor, the International Labour Organization, the German Federal Ministry for Economic Cooperation and Development, and Finmark Trust, created the global "Access to Insurance Initiative" (www.access-to-insurance.org) to foster capacity development and standard setting for insurance supervisors.

Key challenges for the development of microinsurance markets in rural areas

Much has been learned about developing effective and broad-based microinsurance markets in the past few years, but a number of challenges still face efforts to extend microinsurance in rural areas:

1. The strategy and policy challenge: A holistic approach to improving the financial system addresses the actors on three levels: it focuses simultaneously on framework conditions involving sector strategies, regulation, and supervision (macro level); service providers and public goods (meso level); and insurers, intermediaries, and customers (micro level). Rural development policies and financial sector development policies need to include microinsurance, and public resources need to be made available.
2. The underwriting challenge: Underwriting in rural areas faces higher risks and weaker infrastructure, which requires special attention from policymakers and development organizations. Community-based and mutual types of underwriters are more common in rural areas than in urban areas, and they often require institution-building support. International development cooperation agencies can help by supporting capacity building and investment.
3. The delivery challenge: Delivery channels that are close and easy to use and support rural delivery need to be strengthened. These channels can include cell-phone banking and retail shops. Community-based organizations play an important role in providing microinsurance in rural areas, and these organizations are often the risk carriers. This situation is suboptimal because of their limited reserves and management skills. Converting them from insurers to delivery channels could mitigate many of these problems.
4. The consumer challenge: Consumer-related challenges, such as affordability, insurance literacy, and consumer protection, need a special look. Affordability is more sensitive in rural areas than in urban areas because rural residents face higher cash-flow fluctuations in agriculture and generally have lower incomes. More investments are needed in insurance literacy. Consumer protection is a greater challenge in rural areas because the ombudsman and courts are often far away, and a claimant needs to finance travel costs.
5. The support structure challenge: Service providers in microinsurance are often not active in rural areas and not equipped to transfer the required know-how and systems to their clients, the intermediaries, and insurers. Training courses for rural staff of insurers are more costly to organize. Service providers often face higher costs to assess demand or establish risk data for remote areas and therefore neglect rural areas.
6. The agricultural insurance challenge: Although index insurance can potentially overcome many of the problems associated with traditional insurance, it requires improving the availability of high-quality weather data, creating awareness among farmers, achieving quick payouts, and dealing with basis risk due to conditions that might affect farmers but are not incorporated in the index (such as soil composition or uneven terrain).

Conclusion

Rural microinsurance products and their sales strategies require huge investments in product innovation, literacy work, and establishment of sound providers and intermediaries. In addition, regulatory dedication and innovation are required to spur the provision of rural microinsurance by motivating and formalizing rural providers and developing adequate customer protection.

Microinsurance is an integral part of the financial sector and should be promoted as such, through an "access to finance strategy." Rural finance and rural development policies should explicitly deal with microinsurance, including agricultural microinsurance. Coherence with other sector policies—such as agricultural development policy, social security policy, or consumer protection policy—results in more effective approaches. ■

For further reading: German Federal Ministry for Economic Cooperation and Development (BMZ), *Security at Little Cost: Microinsurance in Financial Systems Development* (Bonn, 2009), www.bmz.de/en/service/infothek/fach/konzepte/konzept179.pdf; Lloyd's and the Microinsurance Centre, *Insurance in Developing Countries: Exploring Opportunities in Microinsurance* (London and Appleton, WI, 2009); M. McCord and J. Roth, *Agricultural Microinsurance: Global Practices and Prospects* (Appleton, WI: Microinsurance Centre, 2008), www.microinsurancecentre.org/UI/DocAbstractDetails.aspx?DocID=660.

Martina Wiedmaier-Pfister *(wiedmaier-pfister@t-online.de) is a consultant on financial systems development for German Technical Cooperation (GTZ) and the German Federal Ministry for Economic Cooperation and Development (BMZ).* ***Brigitte Klein*** *(Brigitte.Klein@gtz.de) heads the sector project Financial Systems Development on behalf of BMZ.*

INTERNATIONAL FOOD POLICY RESEARCH INSTITUTE
Supported by the CGIAR
IFPRI® www.ifpri.org

THE WORLD BANK
www.worldbank.org

2020 VISION FOR FOOD, AGRICULTURE, AND THE ENVIRONMENT

Combining Extension Services with Agricultural Credit: The Experience of BASIX India

VIJAY MAHAJAN AND K. VASUMATHI

FOCUS 18 • BRIEF 13 • JULY 2010

India has nearly 90 million farm households. More than 80 percent of these farmers operate on a small or marginal scale, farming less than two hectares of land. They also usually have one or two buffaloes or cows, reared for milk and dung. Most of these small and marginal farmers fall below the poverty line. To reduce overall poverty in India, it is important to enhance the incomes of small and marginal farmers. One way to do that is to provide credit so they can get access to yield-enhancing inputs like seed, fertilizer, and cattle feed, as well as acquire irrigation pumps and crossbred cattle. But these kinds of investments alone will not raise farmers' incomes. Agricultural and livestock development services are also crucial to give farmers knowledge of improved practices and strengthen their links to markets.

BASIX is an Indian livelihood promotion institution working with more than a million poor households. Its mission is to promote sustainable livelihoods for a large number of rural poor people and women. When it started in 1996, BASIX's primary focus was delivering microcredit to its customers. In 2001, however, BASIX asked the Indian Market Research Bureau to carry out an impact assessment, and the results were rather disappointing. Only 52 percent of the customers, who had received at least three rounds of microcredit from BASIX, showed a significant increase in their income (compared with a control group); 25 percent reported no change in income level; and 23 percent reported a decline in their income level. BASIX then carried out a detailed study of those who had experienced no increase or a decline in income and found that the reasons for these results could be grouped into three factors:

1. unmanaged risk;
2. low productivity; and
3. unfavorable terms in input and output market transactions.

This analysis made clear the need for productivity enhancement, risk-mitigation services, and market linkages, as well as the need for rural producers to come together to amass greater bargaining power in the marketplace. In 2002, therefore, BASIX revised its strategy to provide a comprehensive set of livelihood promotion services to poor rural households. This livelihood triad strategy includes provision of financial inclusion services; agricultural, livestock, and enterprise development services; and institutional development services (Table 1).

What services are provided?

Under Agricultural, Livestock, and Enterprise Development (AGLED) services, BASIX currently provides services to farmers growing nine types of crops (cotton, groundnut, soybean, pulses, paddy rice, chilies, vegetables, mushrooms, and lac [a form of organic resin]) and two livestock products (dairy and meat [sheep and goat]). Nonfarm business development services are also provided in selected activities like tailoring, woodworking, bamboo work, and retail stores.

How are services delivered?

BASIX works in more than 25,000 villages through a network of 150 branches, each with five field executives under a team leader. Each field executive supervises five livelihood service advisers (LSAs), who each cover about 10 villages, originating credit, selling insurance, and collecting repayments. The LSAs also sell AGLED Services. BASIX has more than 3,000 LSAs.

BASIX field executives identify and select villages or clusters of villages to receive these services. A cluster is a group of villages within a radius of 6 to 8 kilometers—a size that includes a reasonable customer base for delivering services effectively and efficiently. The branches start enrolling customers for services in those villages where there are at least 30 existing borrowers for either crop or livestock activity.

BASIX has built a cadre of nearly 1,000 livelihood services providers (LSPs). LSAs function as BASIX salespeople, whereas LSPs are similar to extension agents. An LSP works with BASIX on a regular basis and is typically a high-school graduate with training as a para-extension worker or a para-veterinarian. He or she covers 200–400 customers for one crop or activity. BASIX distributes product brochures in regional languages telling customers what

Table 1—Services included in the BASIX livelihood triad

Financial Inclusion Services	Agricultural, Livestock, and Enterprise Development Services	Institutional Development Services
• Savings (directly in districts where BASIX has a banking license, and through other banks elsewhere)	• Productivity enhancement: through increased yields, use of improved seed varieties or practices	• Individual-level awareness building, skill enhancement, and entrepreneurship development
• Credit: agricultural, allied, and nonfarm, short and long term	• Productivity enhancement: reduction in costs	• Formation of groups, federations, and cooperatives of producers
• Insurance for lives and livelihoods, including weather index-based crop insurance	• Risk mitigation (other than insurance), such as livestock vaccinations	• Functional training in accounting and management information systems, using information technology
• Money transfer, for migrant workers	• Local value addition, such as processing cotton into lint before selling	• Building collaboration to deliver a wide range of services
• Experimental products: micropensions, warehouse receipts, etc.	• Alternative market linkages: input supply and output sales	• Sector and policy work: analysis and advocacy for changes and reforms

Source: BASIX.

services they will receive and when. Customers pay Rs 450 (US$10), including services tax, for a year of AGLED services.

Achievements

In 2009 BASIX had nearly half a million customers for AGLED services. About half of these customers were using agriculture and livestock services, and the rest were using services related to nonfarm activities. The details are as follows:

- **Agriculture:** AGLED provided soil-testing service to more than 20,000 farmers, integrated pest management (IPM) or integrated nutrient management (INM) services to nearly 75,000 crop customers, and field surveillance to more than 30,000 farmers. It connected most customers to markets for inputs (seed, fertilizers, pesticides, and bio-inputs such as vermicompost and organic pesticides) and outputs. Weather index-based crop insurance was provided to more than 10,000 farmers for different crops and in different agroclimatic zones, in collaboration with private insurance companies.
- **Livestock:** BASIX AGLED services conducted health checkups of nearly 440,000 animals, vaccinated nearly 165,000 animals, and dewormed 125,000 animals. It trained more than 36,000 customers on feed and fodder and better dairying practices. More than 60,000 farmers were linked to milk marketing chains of cooperatives or private dairy companies. Livestock insurance was provided for more than 120,000 animals, in collaboration with private insurance companies.

Lessons learned

It has taken BASIX about six years to reach the scale described, and it has learned many lessons along the way. In the initial two years, the main emphasis was on market research to identify which services farmers needed. This research, conducted through a large number of field visits and group interactions with farmers, showed that small farmers preferred cost-saving and risk-reducing interventions over yield-enhancing interventions requiring greater cash outlays. It also showed that it was not possible to handle these interventions for a large number of crops, so BASIX focused on a few crops grown by a large number of farmers, such as groundnut in southern Andhra Pradesh, cotton in northern Andhra Pradesh, and soybean in western Madhya Pradesh.

The next step was designing the service offerings. For example, enhancing productivity could mean increasing the yield or reducing the cost for the same output. Local agricultural universities and research stations made available many packages of practices for increasing yields, so BASIX decided to focus more on cost reduction. One successful example of this approach was stem application of pesticide in cotton, which reduced pest multiplication and thereby reduced the need to undertake a large number of pesticide sprays later. Another example was introduction of soil testing, which led to more precise and economical application of fertilizers. In the case of dairy animals, simple practices like vaccination and periodic deworming were more cost-effective than procurement of high-yielding crossbred animals.

Based on such experiences, BASIX staff learned how to customize AGLED services for different agroclimatic zones, which enhanced the farmers' willingness to pay for these services. Customer satisfaction surveys conducted by independent audit teams found that the satisfaction level was nearly 80 percent, with the main cause of dissatisfaction being inadequate visits of the LSPs. To improve service, field executives introduced tighter monitoring of service delivery, but this practice turned out to be expensive. BASIX is now pilot testing mobile phone–based monitoring of service delivery through which farmers will be able to report incidents of no visit or poor service.

Sustainability and future plans

The income from AGLED services in 2009 was nearly Rs 145 million (US$3 million), and BASIX made a modest profit of nearly Rs 22 million (US$450,000) providing these services to nearly half a million customers. With more and more LSPs reaching the breakeven number of customers, profitability is likely to improve. BASIX also plans to move some basic facilities like soil-testing labs and artificial insemination centers under its own control to improve its service to farmers.

Although BASIX agricultural credit operations are aimed at small and marginal farmers, the organization plans to extend AGLED services to larger farmers (to whom BASIX does not extend credit) in the same villages. These farmers' enhanced yields will generate additional output as well as employment opportunities for the landless poor—outcomes that are in line with the BASIX mission. So far BASIX has worked mainly in poorer dryland districts, but it is also considering providing AGLED services in irrigated districts where it has no credit operations. With these changes, BASIX is confident of reaching 2 to 3 million farmers with AGLED services in the next five years. ■

For further reading: BASIX, www.basixindia.com; P. Chandra Shekara, *Status of Private Extension in India* (Hyderabad: National Institute for Management of Agricultural Extension [MANAGE], 2002); R. Sulaiman and V. V. Sadamate, *Privatising Agricultural Extension in India*, Policy Paper 10 (New Delhi: National Centre for Agricultural Economics and Policy Research [NCAP], 2000); S. K. Datta, *A Perspective on Farmer-Market Interface: Results of a Revisit to Selected Villages from West Bengal, Gujarat, and Maharashtra* (Ahmedabad: Indian Institute of Management, 1999); V. Mahajan, "From Micro Credit to Livelihood Finance," *Economic and Political Weekly* 40, no. 41 (2005): 4416–19; V. Mahajan, Beyond Microfinance, in C. Moser, ed., *Reducing Global Poverty* (Washington, DC: Brookings Institution, 2007).

Vijay Mahajan *(vijaymahajan@basixindia.com) is founder and chief executive officer of the BASIX Group.* ***K. Vasumathi*** *(vasumathi@basixindia.com) is associate vice president, Agricultural, Livestock, and Enterprise Development (AGLED) services, BASIX Group. They are based in Hyderabad, India.*

INTERNATIONAL FOOD POLICY RESEARCH INSTITUTE
Supported by the CGIAR
IFPRI® www.ifpri.org

THE WORLD BANK
www.worldbank.org

Bundling Development Services with Agricultural Finance: The Experience of DrumNet

JONATHAN CAMPAIGNE AND TOM RAUSCH

FOCUS 18 • BRIEF 14 • JULY 2010

Agriculture is the largest economic sector in most African countries and remains the best opportunity for economic growth and poverty alleviation on the continent. Yet, sadly, the sector has been in decline over the past 40 years, and poor farmers have largely remained poor. This failure is due to many factors, including collapsed agricultural development banks, corruption, inadequate infrastructure, and poor soils and seeds. It has also occurred because smallholder farmers lack access to critical information, market facilitation, and financial intermediation services.

This brief reviews the DrumNet Project and its approach to improving farmers' access to finance in Kenya. The project has found that financing small-scale farmers is challenging given the cost and risk associated with serving rural, relatively isolated clients. Lending becomes increasingly feasible, however, in a supply-chain approach in which farmers are connected to a formal network of buyers, retailers, and financiers.

The DrumNet Project

Financing farmers is a difficult proposition in Africa. African farmers tend to be geographically dispersed, resource poor, and undereducated—all traits that amplify the costs and risks involved with lending. Other characteristics related to the agricultural sector, such as unpredictable weather patterns, long crop cycles, irregular market access, and volatile or high farm input costs, make the proposition even more unappealing to financial institutions. Consequently, agricultural lending constitutes less than 1 percent of the commercial lending taking place on the continent.

The DrumNet Project has operated in Kenya since 2005 and employs proven microfinance principals and a supply-chain approach to promote agricultural lending (Figure 1). The project establishes relationships with key actors along a supply chain—a buyer, a bank, and several farm input retailers—and links them to smallholder farmers through a dedicated transaction platform and a fully integrated finance, production, delivery, and payment process. The targeted use of information and communication technologies (ICTs) across the platform makes the process efficient, cost-effective, and practical in the African context.

Figure 1—DrumNet farm input loan and payment flow

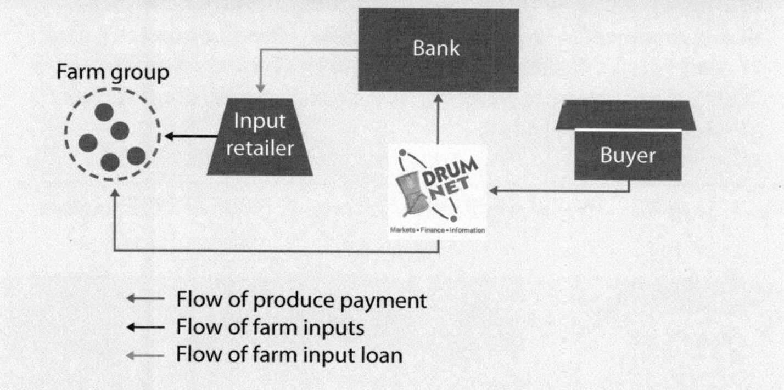

Source: Authors.

The process begins when farmers (organized into farmer groups) sign a fixed-price purchase contract with an agricultural buyer. The contract allows farmers to approach a partner bank, obtain credit, and get farming inputs from a local, certified retailer. At harvest, the contracted produce is collected, graded, and sold to the buyer at designated collection points. A successful transaction triggers a cashless payment through a bank transfer. DrumNet serves as the intermediary in the flow of payment to ensure that credit is repaid before earnings reach farmers' accounts. A master contract governs the entire process, and DrumNet's information technology (IT) system monitors compliance.

The process creates an enabling environment for agricultural finance in several ways. First, banks are assured at the time of lending that farmers have a market for their produce and the means to adequately serve that market—two building blocks of a healthy revenue stream. Second, banks minimize the problem of loan diversion by offering in-kind credit to farmers for inputs and directly paying certified (and monitored) input retailers after distribution of the inputs. Finally, cashless payment through bank transfers reduces strategic default, since farmers cannot obtain revenue until their outstanding loans are fully repaid.

DrumNet has piloted its approach in Kenya's horticultural and oilseed sectors, serving more than 3,000 farmers across five provinces.

Lessons learned, challenges faced

Since its start in 2005, the DrumNet Project has learned many lessons and undergone continual testing and redesign. DrumNet has found that by bundling services at various stages in the supply chain, its approach can enhance efficiency and build trust between actors in the chain. Together, that efficiency and trust help resolve many of the problems that historically discourage smallholder financing.

At the same time, DrumNet has encountered many challenges during implementation, particularly related to partner noncompliance and poor agricultural yields. In both situations, the outcome has been a substantial number of farmer loan defaults that eroded the interest of DrumNet's crop-buying partner and the goodwill of its banking partner. These remaining challenges show that the package of services must be adjusted and enhanced as the project moves forward.

Partner noncompliance

For this approach to function properly, each supply-chain partner must abide by an established set of procedures and rules. Therefore, supply-chain actors must find value in and benefit from the arrangement at all times. In theory, this is the case. Farmers get to produce goods under structured agreements and obtain inputs that help them boost farm productivity. The buyer receives greater quantities of higher-quality product with limited field mobilization. Input retailers realize increased sales without taking on the burden of offering credit themselves. And the bank captures a new customer market with minimal risk, adding to its loan portfolio and deposit base.

DrumNet has, however, experienced its share of noncompliance. Farmers have opted to side-sell produce outside buyer agreements to attain quick cash or evade loan obligations. Buyers have at times failed to honor contract terms, and input retailers have engaged in dishonest practices as well. Even banks have strayed from the program by delaying payments and introducing unexpected fees to farmers.

It is crucial to resolve the issue of partner noncompliance because the benefit any one actor accrues from DrumNet depends on other partners' faithful completion of their functions in an agreed-upon manner. In other words, once one actor fails to comply because he or she does not find value in the arrangement, the overall value of the supply-chain approach is lost.

Low agricultural yields

Farmers' inability to attain sufficient crop yields has also negatively affected the project and its overall service package. In many instances, poor yields have resulted from poor weather conditions. Kenya has experienced several years of irregular and insufficient rain, especially in the eastern portion of the country. Consequently, many farmers have produced only small or extremely stunted harvests.

Soil conditions in Kenya have also diminished farmers' productivity. Because of population pressures, intensified agricultural activities, and low fertilizer use, many Kenyan famers have exhausted their soils. With such soil conditions, even the best agronomic practices result in disappointing yields, low returns on farm investments, and consequently further soil degradation from season to season. The use of poor seed varieties has exacerbated the problem.

The way forward

To address these challenges and others, DrumNet has identified products and services that can be bundled with or added as supplements to the supply chain:

- **Performance rating.** DrumNet is devising a performance rating system that will be integrated into DrumNet processes to allow good and bad performers to be identified, thus creating an incentive for better partner behavior and commitment and distinguishing especially competent, reliable actors over time. Simple credit ratings could also serve as helpful indicators for banks as they assess potential borrowers' creditworthiness.
- **Crop insurance.** A dedicated crop insurance product that insures farmers' inputs against drought or other acts of God would reduce the weather risk inherent in agricultural financing, win further buy-in from farmers, and fill a crucial gap in this bundled, supply-chain approach. The product could be directly tied to input sales or incorporated into production contracts. Farmers would receive not only a guaranteed produce purchase price, but also guaranteed reimbursement or replacement of inputs.
- **Soil analysis.** A soil analysis service would provide farmers with precise recommendations on how best to restore fertility to their soils and, accordingly, improve land productivity. A fertilizer matching component—matching the right fertilizers to a farmer's particular soil composition—would make the analysis even more effective. The analysis could be offered by input retailers, thereby generating greater trust between farmers, retailers, and DrumNet.
- **Payment systems.** Advances in electronic payment options should also play an important role as DrumNet moves into the future. Payment systems like M-PESA, ZAP, and MobiCash can increase the timeliness of transactions between supply-chain partners and move cash points closer to rural-based farmers. These payment solutions, together with the increasing number of bank products available in the market, will reduce the hassles farmers now incur when receiving payment.

DrumNet is now commercializing its operations through the formation of a private company in Kenya. The products and services described in this brief, along with others, are being incorporated into the new company's platform. In addition, the company plans to upgrade DrumNet's existing IT system, building a more robust and expandable system that is accessible to rural-based partners. The system will be modular in structure, so users with different requirements can select and use different components. Furthermore, the system's functionality will be matched by its simplicity—it will easily plug in to the way users already conduct business.

To mitigate risk on a larger scale, comprehensive and commonly accepted standards for communication, financing, information, and exchange must be applied across different agricultural supply chains in Africa. It is in everyone's interest to facilitate and enforce the development of these standards to ensure that all agricultural actors and initiatives operate under a single preferred paradigm. ■

For further reading: X. Giné, "DrumNet Case Study" (World Bank, Washington, DC, 2005), siteresources.worldbank.org/DEC/Resources/DrumnetCaseStudy.pdf; S. Arnquist, "In Rural Africa, a Fertile Market for Mobile Phones, *New York Times*, October 6, 2009, www.nytimes.com/2009/10/06/science/06uganda.html; *The Economist*, "Security for Shillings: Insuring Crops with a Mobile Phone," March 11, 2010, www.economist.com/business-finance/displaystory.cfm?story_id=15663856; *The Economist*, "Dial M for Money: Beating Banks at Their Own Game," June 28, 2007, www.economist.com/business-finance/displaystory.cfm?story_id=9414419.

***Jonathan Campaigne** (jfc@prideafrica.com) is executive director of PRIDE AFRICA. **Tom Rausch** (trausch@prideafrica.com) is PRIDE AFRICA's regional director for East Africa.*

INTERNATIONAL FOOD POLICY RESEARCH INSTITUTE
Supported by the CGIAR
IFPRI® www.ifpri.org

THE WORLD BANK
www.worldbank.org

APPENDIX A: Supplementary Material

Brief 2, "Financial Literacy," by Monique Cohen

Figure 1—Impact framework for financial education

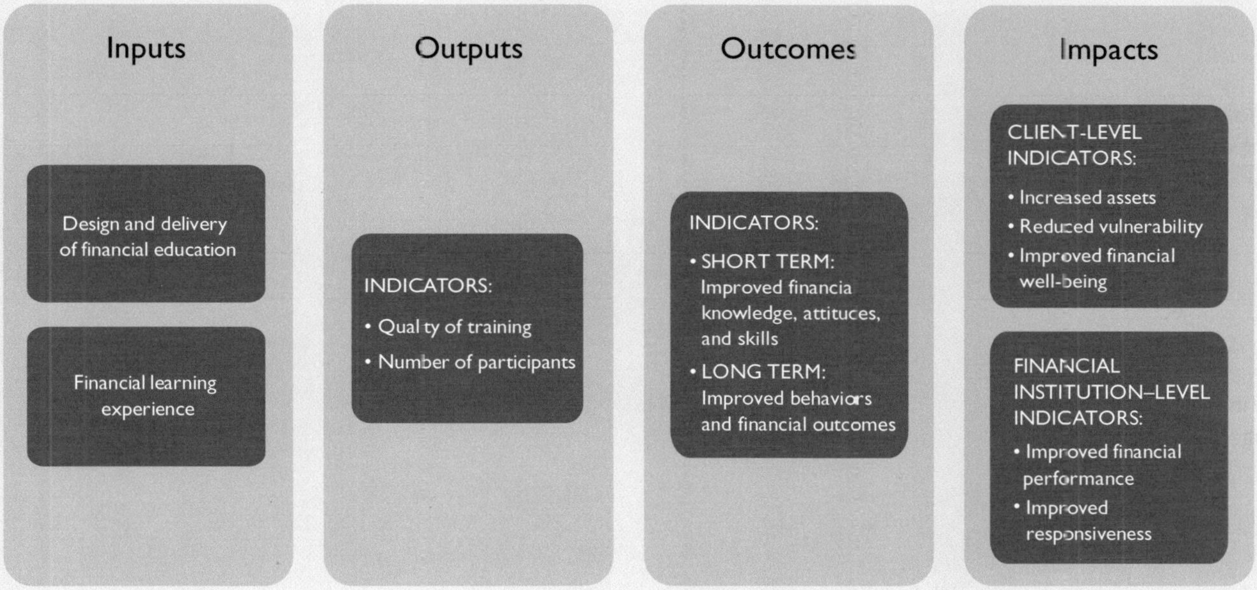

Source: B. Gray et al., "Can Financial Education Change Behavior?: Lessons from Bolivia and Sri Lanka," Working Paper 4 (Microfinance Opportunities: Washington, D.C.: 2010).

Brief 8, "M-PESA: Finding New Ways to Serve the Unbanked in Kenya," by Susie Lonie

Figure 1—FinAccess reports on how people in Kenya send money

First study prior to launch of M-PESA (multiple responses possible). Second study conducted two years post launch.

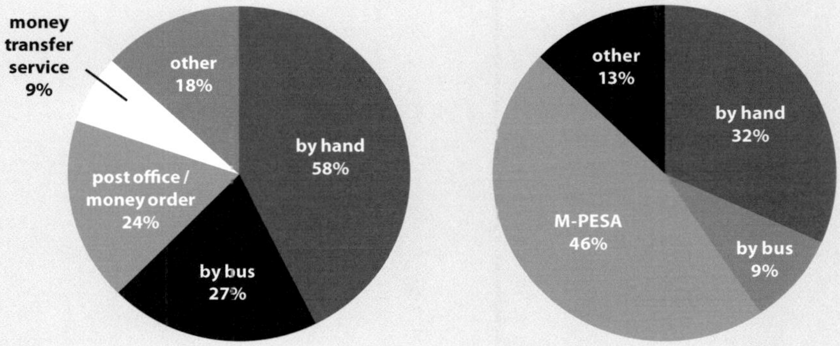

Source: FinAccess 2006

Source: FinAccess 2009

Made in the USA
Monee, IL
05 November 2026

67379791R00021